Transactions on Computer Systems and Networks

Series Editor

Amlan Chakrabarti, Director and Professor, A. K. Choudhury School of Information Technology, Kolkata, West Bengal, India

Transactions on Computer Systems and Networks is a unique series that aims to capture advances in evolution of computer hardware and software systems and progress in computer networks. Computing Systems in present world span from miniature IoT nodes and embedded computing systems to large-scale cloud infrastructures, which necessitates developing systems architecture, storage infrastructure and process management to work at various scales. Present day networking technologies provide pervasive global coverage on a scale and enable multitude of transformative technologies. The new landscape of computing comprises of self-aware autonomous systems, which are built upon a software-hardware collaborative framework. These systems are designed to execute critical and non-critical tasks involving a variety of processing resources like multi-core CPUs, reconfigurable hardware, GPUs and TPUs which are managed through virtualisation, real-time process management and fault-tolerance. While AI, Machine Learning and Deep Learning tasks are predominantly increasing in the application space the computing system research aim towards efficient means of data processing, memory management, real-time task scheduling, scalable, secured and energy aware computing. The paradigm of computer networks also extends it support to this evolving application scenario through various advanced protocols, architectures and services. This series aims to present leading works on advances in theory, design, behaviour and applications in computing systems and networks. The Series accepts research monographs, introductory and advanced textbooks, professional books, reference works, and select conference proceedings.

Jivan S. Parab · Madhusudan Ganuji Lanjewar ·
Marlon Darius Sequeira · Gourish Naik ·
Arman Yusuf Shaikh

Python Programming Recipes for IoT Applications

Springer

Jivan S. Parab
School of Physical and Applied Sciences
Goa University
Taleigao, Goa, India

Madhusudan Ganuji Lanjewar
School of Physical and Applied Sciences
Goa University
Taleigao, Goa, India

Marlon Darius Sequeira
School of Physical and Applied Sciences
Goa University
Taleigao, Goa, India

Gourish Naik
School of Physical and Applied Sciences
Goa University
Taleigao, Goa, India

Arman Yusuf Shaikh
School of Physical and Applied Sciences
Goa University
Taleigao, Goa, India

ISSN 2730-7484 ISSN 2730-7492 (electronic)
Transactions on Computer Systems and Networks
ISBN 978-981-19-9465-4 ISBN 978-981-19-9466-1 (eBook)
https://doi.org/10.1007/978-981-19-9466-1

This Springer imprint is published by the registered company Springer Nature Singapore Pte Ltd.
The registered company address is: 152 Beach Road, #21-01/04 Gateway East, Singapore 189721, Singapore

Foreword

I am extremely happy to write a foreword for the book "Python programming recipes for IoT applications". This book aims to provide a comprehensive guide on how to use Python to develop applications for the Internet of Things (IoT). The IoT has revolutionized the way we interact with the world around us, and Python is one of the most popular programming languages used for developing IoT applications. This book will teach how to use Python to develop solutions that can connect to and interact with IoT devices.

This book is divided into six chapters, each focusing on a different aspect of Python programming for IoT applications. The first chapter will introduce you to the world of IoT and provide an overview of the Python programming language. The subsequent chapters will delve into various topics, such as data handling, networking, security, and machine learning. Each chapter will contain several recipes, which are step-by-step guides that will teach one how to solve specific problems or accomplish specific tasks. The recipes are designed to be easy to follow, and they provide practical examples that one can use to build own IoT applications.

This book is designed for anyone who wants to learn how to use Python to develop IoT applications. Whether you are a beginner or an experienced programmer, this book will provide you with the knowledge and tools you need to build your own IoT solutions. If you are interested in developing applications that can connect to and interact with IoT devices, this book is for you.

In conclusion, I would like to say that Python is an excellent language for developing IoT applications, and this book will provide one with the knowledge and tools one need to get started. Whether you are building a simple sensor network or a complex machine learning application, the recipes in this book will guide you through the process. So, let's dive in and start building some awesome IoT applications with Python.

Taleigao, India
March 2023

Professor Harilal B. Menon
Vice Chancellor
Goa University

The original version of this book was revised. The series editor name was removed in XML of chapters 1, 2 and 6. The correction of the book is available at https://doi.org/10.1007/978-981-19-9466-1_7.

Preface

We are pleased to present this book titled "Python Programming Recipes for IoT Applications" to the readers. As we all know, Python is powerful, yet flexible, easy to learn, and can be adapted to work with most of the microcontroller-based environments. Simplicity serves as a great asset to Python, which allows it to flourish on every platform. The coding flexibility and dynamic nature of Python help developers to create intelligent Internet of Things (IoT) devices.

So, we thought this is the right time to bring such a book to the market which explains the basic glimpses of Python followed by three different embedded platforms and their configuration setup. Various IoT applications with FoG and cloud-based computation on these platforms are explained in a very simple and understandable manner so that the readers can start programming on any of the platforms of their choice with ease.

The detailed chapter-wise flow is as follows:

Chapter 1 "Python Programming and IoT" discusses the basics of Python language such as features of Python, Integrated Development Environment (IDE), data types, and so on. It also discusses comparison of Python with C and C++ language. The C/C++ programming languages dominates embedded systems programming. On the other hand, Python has many strengths that make it a great programming language for embedded systems. This chapter also gives the overview of IoT and its applications.

Chapter 2 provides detailed step-wise configuration setup of Raspberry Pi, MicroPython Pyboard, and NVIDIA Jetson Nano This chapter also covers the details of all above boards including block diagram, features available in these boards, and functions of each block.

Chapter 3 gives the detailed implementation steps of simple IoT applications using Raspberry Pi such as controlling LED blinking, OLED display interface, camera interface, and motor control (DC motor, stepper motor, and servo motor).

Chapter 4 gives the detailed implementation steps of MicroPython Pyboard for IoT applications such as home automation, smart e-waste bin, industrial environment monitoring, green house monitoring, and aquaculture monitoring.

Chapter 5 focusses on FoG and cloud computing with NVIDIA Jetson Nano board. It covers the introduction and the model architecture of both the FoG and cloud computing. Detailed stepwise implementation of patient monitoring with cloud computing and home security (home surveillance, home safety lock, fire alert system) with FoG computing.

Chapter 6 covers the implementation of Machine Learning (ML) applications such as pattern recognition, object classification, and prediction using Jetson Nano.

Taleigao, India

Jivan S. Parab
Madhusudan Ganuji Lanjewar
Marlon Darius Sequeira
Gourish Naik
Arman Yusuf Shaikh

Acknowledgement

The book is a result of practical implementation of various IoT applications using Python programming on Raspberry Pi, MicroPython Pyboard, and NVIDIA Jetson Board with a hands-on approach. Special thanks to Mrs. Kamiya Khatter, Associate Editor and publishing team of Springer to bring this book in the market.

Authors are thankful to Prof. Harilal. B. Menon,Vice-Chancellor and Prof. Vishnu Nadkarni, Registrar, Goa University for encouraging and providing administrative and financial support. The main motivational force behind this book is Professor, Dr. Rajanish Kamat (Vice-Chancellor of Dr. Homi baba State University, Mumbai) who has co-authored several books with authors.

We are thankful to Prof. Kaustubh Priolkar, Dean of School of Physical and Applied sciences (SPAS), Prof. Ramesh Pai, Vice-Dean (Academic), and Prof. Rajendra Gad, Vice-Dean (Research) of SPAS.

Authors would like to acknowledge with gratitude, the support and love of family members. They all kept us going and this book would not have been possible without them. Our sincere gratitude goes to Mr. Saish S. Nayak Dalal and Mr. Sameer Patil for helping us in editing the book.

Finally, authors wish to express their gratitude to God, whose presence has given them strength to finish the book.

We will be failing in our duties, if we do not mention the support, encouragement received from friends, supporting staff and colleagues.

Contents

About the Authors

Jivan S. Parab is an Associate Professor & Programme Director of the School of Physical and Applied Sciences, Goa University, India. He has completed his Ph.D. from Goa University in 2011 with the title "Development of novel Embedded DSP architecture for Non-Invasive Blood Glucose analysis". He received his M.Sc. and B.Sc. degrees in Electronics from Goa University, in 2005 and 2003, respectively. He has co-authored four books and several papers in national and international journals and conferences. He has been awarded the Visvesvaraya Young Faculty award of Rs. 38 lakhs by MeiTy, the Government of India. He has completed two industry consultancies and two major research projects. He also has three Indian patents to his credit. His research interest is Embedded system design, Signal processing, Machine Learning, Biomedical Instrumentation, and Agro-Instrumentation.

Madhusudan Ganuji Lanjewar is a Technical Officer at the University Science Instrumentation Centre (USIC), School of Physical and Applied Sciences, Goa University, India. He received his M.Sc. Electronics in 2003 and B.Sc. Electronics degree in 2001 from RTM Nagpur University. His research interests include artificial neural networks, image processing, and embedded system using the "C" language and Python. He has published several papers in national and international level journals and conferences. He has two patents to his credit.

Marlon Darius Sequeira is an Assistant Professor of Electronics at the School of Physical and Applied Sciences, Goa University, India. He received his B.Sc. and M.Sc. degrees from Goa University in 2012 and 2014 respectively. He obtained his Ph.D. from Goa University in 2021. His research interests include biomedical Instrumentation, machine learning, embedded systems, and FPGA designs. He has published several papers in national and international journals and conferences. He has two patents to his credit.

Gourish Naik Former Dean Faculty of Natural Science & Former Head Department of Electronics obtained his Ph.D. from the Indian Institute of Science, Bangalore (1987) and served the institute as a research associate in the areas of Optoelectronics and Communication until 1993. He was associated with the Goa University for 27 years. He is the founding head of the University Instrumentation Centre and established fiber optic LAN and wireless communication networks at Goa University. He was also the coordinator of DEITI (an educational broadcast studio supported by Indian Space Research). His other commitments are regulating the digitization center at Goa University to support the various digital repository projects like DIGITAP (Digital Repository for Fighter Aircrafts Documentation) of the Indian Navy, the Million Book project of the Ministry of Information Technology and Antarctica Study Center (NCAOR). He has to his credit around 60 research papers published in international journals and has presented research works at various national and international forums. He has delivered several keynote addresses and has been invited to talks at various institutes and also authored five books on embedded systems and allied areas published by CRC Press, Springer, Lambert, etc. He was a member of the Goa State Rural Development Authority and also an advisor for the Directorate of Education. He was a governing body member of the engineering college of Goa and also a member of the faculty board of Goa University.

Arman Yusuf Shaikh is currently associated with the industry-leading VLSI company as an ASIC Design engineer, where he is responsible for taking a design through RTL-GDS2 flow involving steps like Synthesis, Floor planning, PNR, STA, Et Cetera. He is a motivated professional holding an M.Sc. in Electronics degree from Goa University (2021) and a B.Sc. in Electronics from Dnyanprassarak Mandal's College and Research Center (2019). As a Research Student at SPAS Electronics, Goa University, Arman completed a couple of projects that exposed him to the domains of Biomedical Instrumentation, ML, IOT, Signal Processing, and Embedded systems. He has published papers in notable journals and conferences. He has two patents to his credit.

Chapter 1
PYTHON Programming and IoT

Abstract There are number of books that cover Python programming in detail. This chapter is not exhaustive; instead, it briefly describes Python's programming basics that will give a good hold for programmers to write the programs. It answers the question of whether Python will replace C/C++? This chapter also briefly introduces IoT and Python for embedded and IoT applications.

Keywords Python · IDE · Operators · Internet of Things · IoT applications · Embedded system

1.1 Introduction to Python

Python is a widely used general-purpose interpreted programming language that allows to work more quickly and integrates the systems more perfectly. It is easy to learn, supports object-oriented programming with dynamic semantics which enables a programmer to write intelligible and logical code for small and bigger entities (Kuhlman 2012). Due to the availability of extensive standard libraries, it is considered as "batteries included" language.

Why the Name Python?

The first question that arises in a programmers' mind is that why the name Python is given to this interesting programming language. Does this have some relation with the Python Snake? There is one interesting fact behind choosing the name Python. In the 1980s, Guido Rossum developed Python and wanted to choose a distinctive name and a little bit funny one too. While reading the interesting script of a British Broadcasting Corporation (BBC) comedy series, which was quite popular i.e. "Monty Python's Flying Circus". So, he made his mind to name their programming language as "Python".

The original version of this chapter has been revised. Series editor name has been removed from online version of this chapter. The correction to this chapter is available at https://doi.org/10.1007/978-981-19-9466-1_7

J. S. Parab et al., *Python Programming Recipes for IoT Applications*, Transactions on Computer Systems and Networks, https://doi.org/10.1007/978-981-19-9466-1_1

History and Different Python Versions

The foundation of Python was laid in the late 1980s but, practical implementation was commenced in December 1989 by Guido at CWI in the Netherlands. Today Python has evolved as such a popular programming language with several releases, and with every release, some new features are added in it.

Python versions:

- The first Python code was published with labeled version 0.9.0 by Guido Van Rossum in February 1991.
- Python 1.0 was manumitted in 1994 with features such as lambda, map, filter, and reduce, followed by Python 1.5 and 1.6 in the years 1997 and 2000 respectively.
- Python 2.0 was released with addon features such as comprehensions, garbage collection systems. Python 2 also has intermediate version 2.1 till 2.7.
- Similarly, on December 3, 2008, Python 3.0 (also called "Py3K") was manumitted. Python 3 was specially developed to rectify the basic flaws of the earlier versions. From 2008 till date several releases of Python 3 have come, from version 3.1 till 3.11 which was released in February 2023.

1.2 Can Python Replace C/C++?

The answer to the question "can Python replace C/C++?" is **NO**. C and C++ forms the basis of every programming. Python is built using C by keeping web programming in mind. So there is no possibility that Python will replace fundamental languages like C or C++ at least not in the near future.

Overall Python is better than C/C++ in terms of its simplicity, readability, writability, and very simple syntax, while C/C++ is tedious and time-consuming, prone to errors, and not easy to understand . Although C/C++ is better in terms of performance, speed and has enormous application areas.

Python offers enormous error-checking option than C. Being a very high-level programming language, it has various data types, arrays, and dictionaries already built-in that would have cost someone many days to implement effectively in C. Python gives the flexibility to break up the program into small modules which later can be leveraged in another Python programs.

For embedded system code development, the programmer prefers C/C++ language due to faster runtime code. When it comes to runtime speed Python may be less efficient than C/C++, however, runtime speed is sometimes not that important when development speed is considered, which is good for Python. Now in recent times, Python is extensively used for embedded systems explained in Sect. 1.4.

1.3 Overview of Python Programming

Python programming language is easy to learn because of its simple syntax and huge library packages. Before starting with the basics of Python programming one has to understand the process of Python and Integrated Development Environment (IDE) installation to get acquainted with the respective IDE's environment.

1.3.1 Python Installation

Python official website (https://www.Python.org/) provides the most updated and current source code, binaries, documentation, news, etc. Python can be downloaded from the above website link. The Python can also be installed with the Anaconda . In order to install Python on a windows machine one has to follow the below steps.

Installation steps on Windows machine

Step 1: Visit a web page https://www.Python.org/downloads/, as shown in Fig. 1.1. Click on the 'Download' button to download the latest Python version. If someone wants the older versions of Python, the same is available for download on the same link.

Fig. 1.1 Python downloading web portal

Step 2: The Python version 3.9.7 is now downloaded and saved in the Download directory. Search file name Python 3.9.7-amd64 → double click on it → click on 'Run' and follow the pop up instructions till it shows the finished installation.

Anaconda Installation

Anaconda is an incredible collection of Python packages, tools, resources, and IDEs. This package comprises several tools which are helpful for the data scientist to exploit Python's unbelievable strength. Anaconda is a free and open-source individual edition available for developers. The installation steps are given below:

Step 1: Visit the web page https://www.anaconda.com/products/individual-dto and download the individual edition of the Anaconda. Click on 'Download' button as shown in Fig. 1.2a.

Step 2: Search.exe file (Anaconda3-2021.05-Windows-x86_64) in Download directory. Double click on it → click on 'Run' → click on 'Next' (Fig. 1.2b) to continue → click on 'Agree' (Fig. 1.2c) → select 'Just me' and then 'Next' (Fig. 1.2d) → choose the location for installation (Fig. 1.2e) and click on 'Next' → click on 'Register Anaconda3 as my default Python' (Fig. 1.2f) → click on 'Install' button.

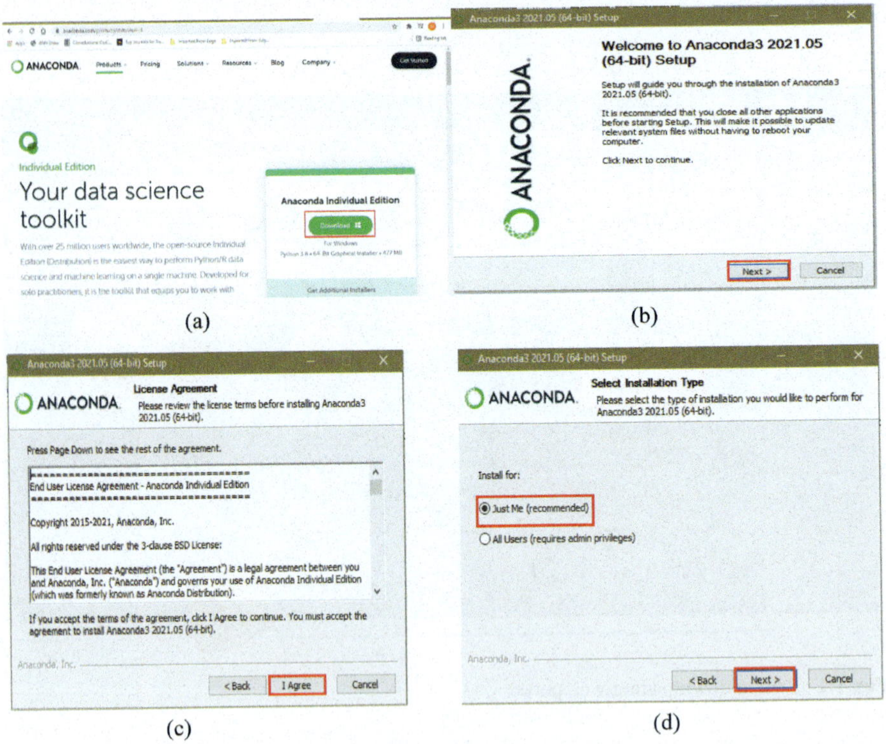

Fig. 1.2 Anaconda installation

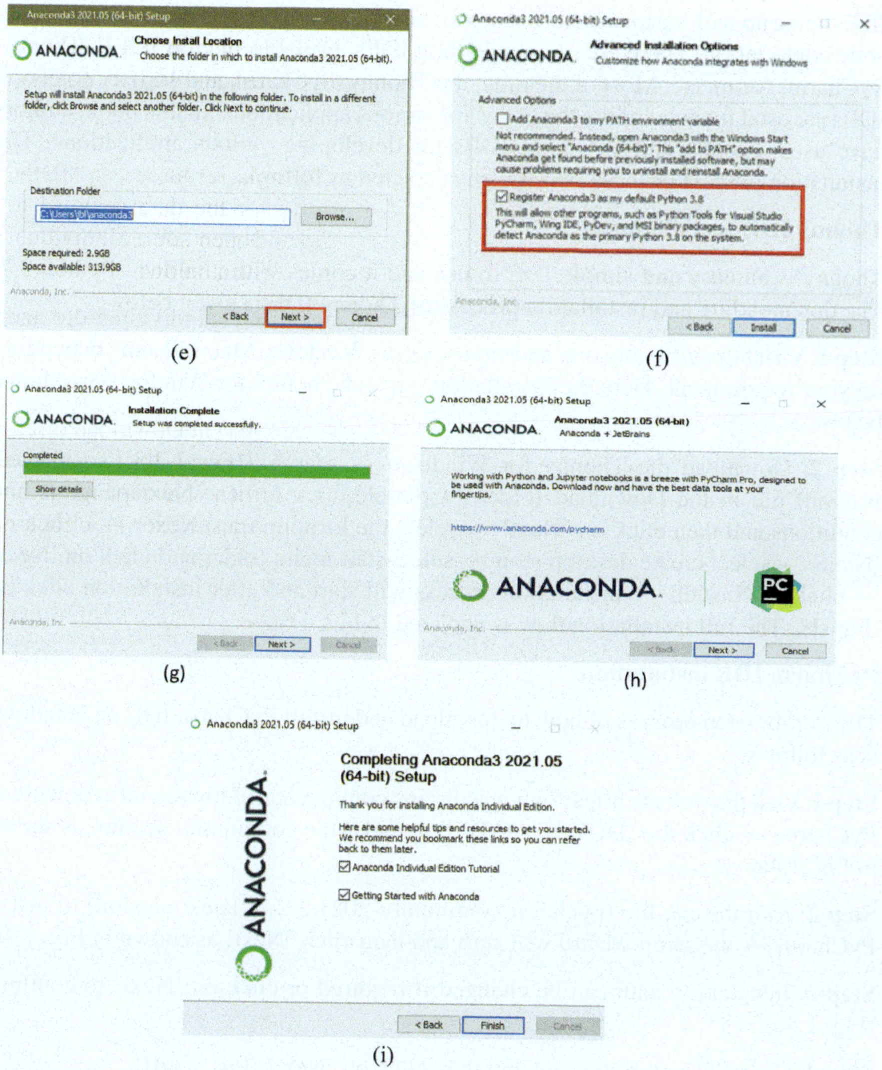

Fig. 1.2 (continued)

Step 3:After completion of installation process (Fig. 1.2g) click on 'Next' → click on 'Next' button to finish the process as shown in Fig. 1.2h and then 'Finish' button (Fig. 1.2i).

Python IDEs and Code Editors

A tool for writing and editing code is called a code editor. Usually, code editors are lightweight and easy to learn. However, if the program is large, then IDEs are helpful to understand the program in a much better way compared to text editors.

IDEs come up with various features such as, build automation, testing, debugging, and code compilation. There are various Python IDEs available such as IDLE, Thonny, PyCharm, Atom, etc. Most of the time, the Thonny, PyCharm, and Jupyter notebook IDEs are used for developing the codes for various applications. In this book, authors have used Thonny and PyCharm IDEs for developing various applications. The installation process of these two IDEs is depicted as follows.

Thonny IDE Installation

Thonny is an easy and simple IDE to use and it comes with a built-in Python 3.7. The downloading and installation process of Thonny IDE is given below:

Step 1: Visit https://thonny.org/ and choose OS as Windows, Mac, or Linux depending on your requirement. Here the installation steps of Thonny for Windows are shown below.

Step 2: Download the Thonny for Windows → search Thonny-3.3.13 (or latest version) file in the Download folder → double click on it → accept terms and conditions and then click on 'Next' → select the location (destination) → click on 'Next' → select create desktop icon → select start menu folder and click on 'Next' → click on 'Install' → installation process will start and after installation click on 'Finish'. The full installation flow is depicted in Fig. 1.3a–e.

PyCharm IDE installation

The step-by-step process of how to download and install PyCharm IDE on Windows is as follows:

Step 1: Visit the website https://www.jetbrains.com/pycharm/download/ to download PyCharm → click the 'DOWNLOAD' link under the community section as shown in Fig. 1.4a.

Step 2: Run the.exe file (pycharm-community-2021.2.2 or latest version) to install PyCharm → the setup wizard will start and then click 'Next' as shown in Fig. 1.4b.

Step 3: Installation path can be changed if required or click on 'Next' to continue (Fig. 1.4c).

Step 4: Create Desktop shortcut and then click on 'Next' (Fig. 1.4d).

Step 5: Choose the start menu folder (Fig. 1.4e) → keep JetBrains selected → click on 'Install' → wait until the installation process is finished.

Step 6: After finishing the installation, a message screen will display as shown in Fig. 1.4f → click on 'Finish'.

1.3.2 Glimpses of Python Programming

Python is a fantastic language for beginners and helps to develop various applications. Following are some of the important characteristics which make this language very powerful:

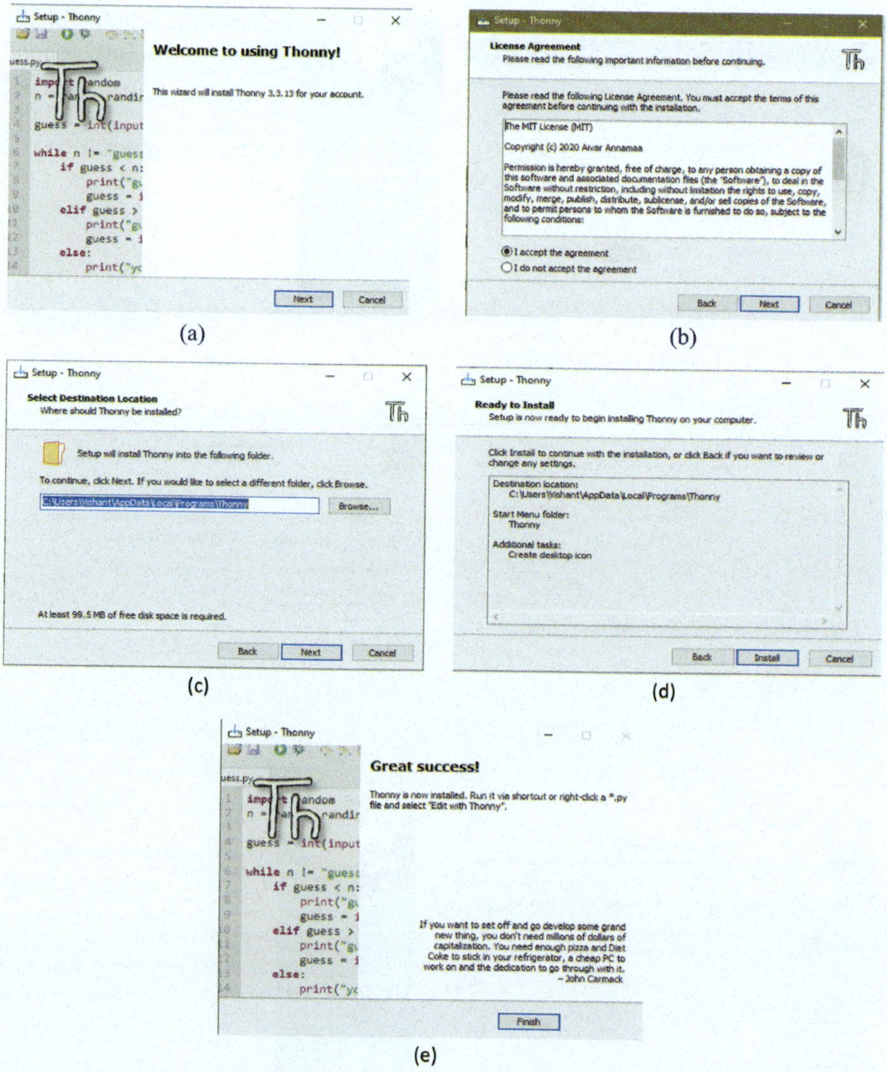

Fig. 1.3 Thonny installation process

- Python supports Object-Oriented Programming (OOP) as well as it is structured programming.
- It supports scripting language.
- It accepts dynamic data types and does dynamic type checking.
- It can be efficiently encapsulated using C, C++, COM, ActiveX, CORBA, and Java.

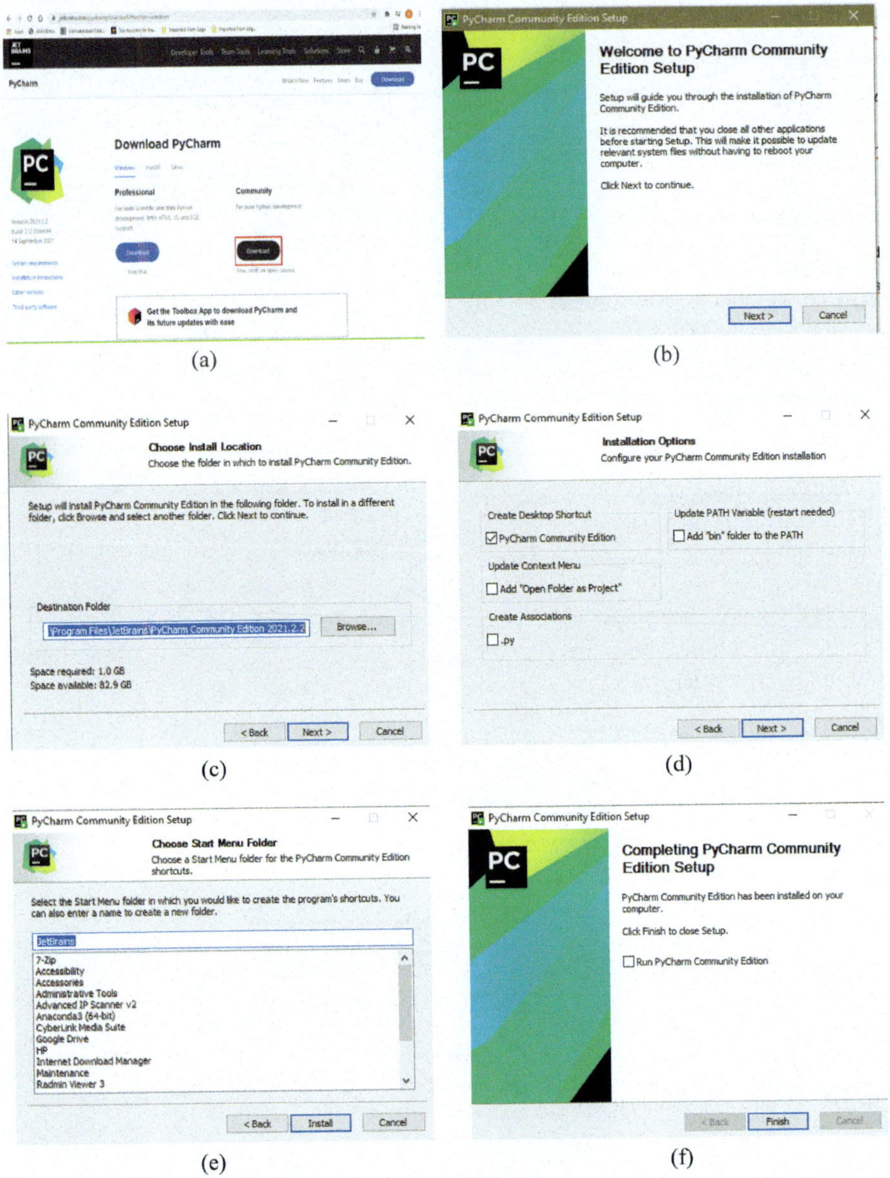

Fig. 1.4 PyCharm installation steps

Table 1.1 Keywords in python

And	Assert	Break	Class	Continue	Def	Del	Elif	Else	Except
Exec	Finally	For	From	Global	If	Import	In	Is	Lambda
Not	Or	Pass	Print	Raise	Return	Try	While	With	Yield

Syntax of the Python Programming Language

This section outlines the general syntactical elements of the Python language, but not in an exhaustive way. The rules on how to use identifiers, reserved keywords, lines and indentations, quotations, different data types, operators, functions, etc. are outlined. After navigating through the following section, the reader will be able to write their first Python code for any desired applications.

Identifiers

An identifier refers to the name used to represent a variable, function, module, or class in Python. It must always begin with an uppercase letter (A–Z) or lowercase letter (a–z) or an underscore, and then can continue with a letter, underscore, and digits (0–9). We cannot use special characters such as @, $, and % in building an identifier. It must be noted that Python is a case-sensitive language and care must be taken to construct the identifier accordingly. For example, **var** and **VAR** are considered two separate identifiers in Python.

Reserved Words

Reserved words cannot be used as identifiers or variables or constants. The Python keywords are always constructed using lowercase letters. Table 1.1 gives the keywords which are used in the Python language.

Indentation in Python

Unlike other languages, a block of code under functions and flow control in Python is not specified using braces or brackets. Instead, Python relies on line indentations, which are rigidly applied in the language.

One can use a variable number of spaces to specify indentations, but this number must be consistent in the entire block. An example showing proper indentation is shown below:

```
a=10
b= 10
if (a==b):
  print("Yes")
else:
  print("No")
```

Multi-line Statements

Usually a statement is terminated by a new line in Python. However Python permits to continue line by using a line continuation character(\):

```
history=10
geography=9
maths=2
science=3
Marks = history + geography +maths + science
print(Marks)
```

However, statement which are contained in {}, [], or () do not require the line continuation character.

```
Employees= ['John', 'kevin', 'Wenzel', 'Albert', 'Peter']
```

In Python, one can use the semicolon (;) character to type multiple statements on the same line provided neither of the statement starts a new code block.

```
print("hello"); a = 50; a = a + 34
```

Quotation

A string in Python is specified using quotes. Single, double, or triple quotes are used to specify literals provided the same type of quotes are used to terminate the string. However, triple double quotes allow spanning a string across multiple lines.

```
name = 'Haston'
message= "Hello World of Python programming"
long_message = """We are going to land on the island tomorrow after
the sunset"""
```

Comments

In Python, a comment is indicated to the interpreter by a hash sign (#) and this character must not be contained in a string literal. The characters placed after the # till the end of the line makes up the comment. The Python interpreter will ignore all the characters after # character.

```
# This line is a comment, below code prints hello Python on the screen
print("hello world") # command to print
```

The Python interpreter ignores the triple single quoted string, as such can be used as a multiline comments.

```
'''Comment can span multiple Lines '''
```

1.3.3 Data Types in Python

Different data types are featured in Python namely numerical and alphanumeric The standard data types allowed in Python are boolean, numeric, strings, lists, tuples, dictionaries, and sets etc. (Beazley and Jones 2013; Pilgrim and Willison 2009). Figure 1.5 shows the various data types in Python.

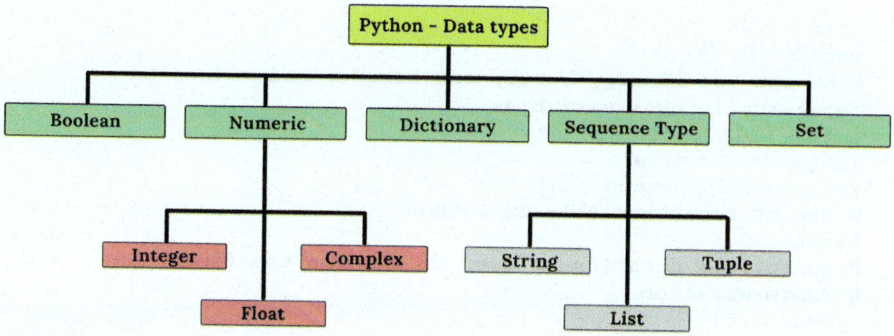

Fig. 1.5 Data types in Python

Boolean

The boolean type is a built-in data type in Python which is used to store only two values, namely, 'True' and 'False'. The main use of the boolean type is to represent the 'True' value of an expression. The code below prints the type of the variable z which is <class 'bool'>:

```
z = True
print(type(z))
```

Numeric

Numeric data types are one of the most extensively used data types in Python. It is further subdivided into integers, float, and complex types. Integers can be positive, negative, and whole numbers including zero which do not have a fractional part and have unlimited precision. Integers can be specified with binary, octal, and hexadecimal values. The numbers 78, -58, and 0 are examples of valid integers in Python. It must be noted that leading zeros are not allowed in non-zero numbers.

Floating-point data type are positive and negative real numbers that have a fractional part. The fractional part is specified after the decimal point. One can use a decimal point representation or scientific representation to specify a floating-point number. For example, 1000.47 is decimal point representation and 1.00047E3 is in scientific representation. The scientific representation is particularly useful to show numbers having many digits.

The third type of number which Python supports is the Complex number. It has a real part and the imaginary part. The imaginary part of the number is multiplied by j (j refer to $\sqrt{-1}$). For example, for the complex number $3 + 2j$, the 3 is the real part and the 2j is the imaginary part. Care must be taken to use j or J as the imaginary component, the use of any other character results in a syntax error. The code below shows the use of integer, float, and complex numbers:

```
# The numerical types in Python
# Use of integer data type in Python
int1 = 15
```

```
int2 = 2500
int3 = -25
# We can use binary, octal, and hexadecimal values to
represent integer variables
int4 = 0b11011011 # binary
int5 = 0o12 # octal
int6 = 0x12 # hexadecimal
# Use of float variable in Python
float1=10.5
# Scientific notation is used to shorten the float
# representation
float2 =1e4
float3 =3.664547e2
# Use of complex data type in Python
comp= 10 + 4j
```

Python Dictionary

Dictionary in Python is similar to a hash table. They operate similar to hashes or associative arrays which are part of the Perl programming language. Dictionary is a pair of key and value. Any valid type in Python can be specified as a key, however numbers and strings are preferred over others. Whereas any arbitrary Python object can be a value for the key.

It must be noted that curly braces ({ }) are used to indicate a dictionary whereas square braces ([]) are used to assign and access.

```
dictionary={}
dictionary['one']="Test one"
dictionary [2] ="test two"
smalldictionary={'name':'tony','srno':789,'section':'research'}
print(dictionary['one']) # Displays value of key 'one'
print(dictionary [2]) # Displays value of key 2
print(smalldictionary) # Displays entire dictionary
print(smalldictionary.keys()) # Displays the keys
print(smalldictionary.values())# Displays the values
```

Sequence Type

Sequence type in Python refers to an ordered set. Strings, tuples, and lists are the popular sequence data types used in Python.

Python Strings

A string is a set of characters that are specified within quotation marks. We can use single or double quotes as a pair to contain a string. Python allows to access a subset of string using the slice operator [] or [:]. This needs to be done using the index which begins at 0. The '+' operator is used to concatenate two strings, and the '*' operator is used to repeat the string as shown in below example:

```
string ='Hello Python !'
print(string) # Displays whole string
print(string[0]) # Displays first character in string
```

```
print(string[1:8]) # Displays 1st to 7th index characters
print(string[3:]) # starting from 3rd index is Displayed
```

After executing above code the output is shown below: *Hello Python ! H ello Py lo Python !* The index can begin with 0 and end with (length -1). The use of concatenate is shown below:

```
print(string *3) # Displays string 3 times
print(string +" World") # concatenated string is Displayed
```

Python Lists

Python has powerful compound data types, one of them is the List data type. The items within the list are specified by a comma separator. All the items are contained within square brackets. A Python list allows entering different data types within the list in contrast to arrays in other programming languages.

```
list =['albert', 145, 3.14, 'raman', 36]
smalllist =[654, 'bose']
print(list) # Shows the list
print(list[0]) # Shows only the the 1st element of the list
print(list[0:4]) # Shows elements starting from 1st till 4th
print(list[2:]) # Shows elements starting from 3rd till end
print(smalllist*4) # Shows the list 4 times
print(list + smalllist) # Prints the concatenated list
```

Python Tuples

Yet another powerful data type in Python is a tuple. This again stores a sequence of data and is similar to a list in its operation. In a tuple, commas are used to separate the items contained within it. However, as compared to a list, the items within a tuple are enclosed within parentheses.

Here we summarize the main differences between a list and a tuple. Square brackets ([]) are used to enclose items in a list whereas parentheses (( )) are used to enclose items in a tuple. In the case of a list, we can change the elements and size, but in a tuple, we cannot change the elements and size. The tuple is a read-only list.

```
tuple =('albert',145,3.14,'raman',36)
smalltuple =(654,'bose')
print(tuple) # prints the tuple
print(tuple[0]) # Displays only the the1st element of the tuple
print(tuple[0:4]) # elements are Displayed from 1st till 4th
print(tuple[2:]) # Displays elements from 3rdtill end
print(smalltuple*4) # Displays the smalltuple 4 times
print(tuple + smalltuple) # Displays the concatenated tuple and
smalltuple
```

Set

Python programming language enables us to use the concept of set. A set is defined as a collection of items that are not in order. The items must be enclosed in curly braces ({ }) and commas are used to separate the items. The interpreter automatically removes duplicate items. The basic operations such as set intersection, union, difference, and symmetric difference are allowed in Python. Indexing is not allowed for a set as the items are not ordered.

```
a = {4,5,5,6,6,6} # a set with repeated values,
b = {4,7,8} # set b
c = a.union(b) # set c is union of a and b print (c)
d = a.intersection(b) # set d is intersection of a and b print (d)
e = a.difference(b) # set e is difference of a from b print(e)
f = a.symmetric_difference(b) # set f is symmetric difference
print(f)
```

Variable Assignment in Python

As the variable is instantiated, the Python interpreter allots memory locations exclusively reserved to that variable. The memory locations allocated to the variable will depend on the data type and dictate what type can be stored in it. Decimals, integers, and characters can be stored in the variables.

The declaration in Python is not explicit and is handled automatically when a value is assigned to the variable. The variable assignment is done as shown in the below examples using the assignment operator (=). The value on the right of the operator is assigned to the variable on the left of the operator.

```
speed = 1000          # An assignment of integer
pi = 3.14             # an assignment of floating-point number
name = "Albert"       # string assignment
print(speed)
print(pi)
print(name)
```

The above lines of code explain how to assign values to variables. The value 1000, 3.14, and "Albert" are assigned to the variables speed, pi, and name, respectively and then printed.

We can also handle multiple assignments in the Python language. In the statement below, the variables j, k, and l are initialized to a value 15.

```
j=k=l=15
```

To perform multiple object assignments to multiple variables the below statement can be used:

```
x,y,z= 21,22, "Albert"
```

An integer object having values 21 and 22 are assigned to x and y, respectively, and at the same time variable z is initialized with string object "Albert".

1.3.4 Python—Basic Operators

Operators are essential constructs of any programming language which are used to manipulate the value of operands. In the expression: 7 * 3, 7 and 3 are the operands and '*' is the operator. In Python, operators are classified into various types which are intended for specific purposes. The different categories of the operators are listed below:

- Arithmetic.
- Comparison (Relational).
- Assignment.
- Logical.
- Bitwise.
- Membership.
- Identity.

Arithmetic Operators

The arithmetic operators are used to perform various arithmetic operations on the operands. Table 1.2 explains all the arithmetic operators assuming that the variables x and y are the operands on whom the operators are applied upon.

Comparison Operators

Operators which compare the operands on each of its sides and find the relation between them are comparison operators. Another name for comparison operators is relational operators. Table 1.3 gives the different comparison operators.

Assignment Operators

These operators are used to assign a value to the variable on the left of the command. Different assignment operators are given in Table 1.4.

Bitwise Operators

Bitwise operators perform bit-by-bit operations. If we assume that x = 126 and y = 31. Then their binary representation will be 0111 1110 and 0001 1111, respectively.

The bitwise operators in Python take the two operands and perform the desired operation on every corresponding bit of two operands. For bitwise AND, it takes the two operands and performs a bitwise AND operation. For bitwise OR, it takes

Table 1.2 Python arithmetic operators

Operator	Code	Description
+ (Addition)	x + y	Addition is performed on operands x and y
− (Subtraction)	x–y	Subtracts right operand y from the left operand x
* (Multiplication)	x*y	Multiply left operand x with the right operand y
/ (Division)	x/y	Left operand x is divided by the right operand y
% (Modulus)	x%y	Left operand x is divided by right operand y and assigns its remainder
** (Exponent)	x**y	Finds the left operand x raised to the power of right operand y
// (Floor division)	x//y	Does the Floor Division of the operand on the left x with the operand on the right which is y

Table 1.3 Python comparison operators

Operator	Python command	Description
==	(x == y)	The result of the operation will be True if the operands x and y are equal
!=	(x != y)	The result of the operation will be True if the operands x and y are not equal
<>	(x <> y)	The result of the operation will be True if the operands x and y are not equal
>	(x > y)	The result of the operation will be True if the operand x is greater than operand y
<	(x < y)	The result is True if the operand x is less than operand y
>=	(x >= y)	The result is True if the operand x is greater than or equal to operand y
<=	(x <= y)	The result is True if the operand x is less than or equal to operand y

the two operands and performs a bitwise OR. Similarly, for bitwise XOR, it takes the two operands and performs a bitwise XOR. In a bitwise complement, it takes the operand and performs a bitwise compliment. In addition, there are bitwise shift left and shift right operators. In the left shift, the bits of the operand on the left are shifted left by the value of the right operand. Similarly, in the right shift, the bits of the operand on the left are shifted right by the value of the right operand. Table 1.5 gives the different bitwise operators.

Table 1.4 Python assignment operators:

Operator	Python command	Description
=	x = y	The value of the right-side operand is assigned to the left-side operand
+=	x += y	The value of right-side operand is added to the operand on the left side and then assigned to the left-side operand
-=	x -= y	The value on the right side of the operator is subtracted from the value on the left side and then assigned to the left-side operand
*=	x *= y	The value on the right side of the operator is multiplied with the value on the left side and assigned to the left-side operand
/=	x =/ y	The value on the left side of the operator is divided with the value on the right side and the result is assigned to the left operand
%=	x %= y	The modulus of the two operands on either side is taken and the result is assigned to the left operand
**=	x**= y	After performing the power operation on the two operands the value is assigned to the left operand
//=	x//= y	After performing the floor division operation on the two operands the value is assigned to the left operand

Table 1.5 Python bitwise operators

Operator	Command	Result
& (bitwise AND)	x&y	0001 1110
| (bitwise OR)	x|y	0111 1111
^ (bitwise XOR)	x^y	0110 0001
~ (bitwise compliment)	~ x	1000 0001
<< (bitwise Left Shift)	x << 2	1111 1000
>> (bitwise Right Shift)	x >> 2	0001 1111

Logical Operators

There are three logical operators, namely, NOT, AND, and OR which are supported by Python. These are used to make a logical decision.

In logical AND, if both the operands are true or non-zero, then only the result will be True. In logical OR, if either of the operands are True or non-zero then the output is True. In logical NOT, the output is the reversal of the logical state of the original operand. Table 1.6 gives the logical operators in Python.

Membership Operators

A membership operator is used in Python to check if the element is a member of a sequence, such as tuples, lists, or strings. The 'in' operator is used for this purpose and will evaluate to True if it finds a variable in the mentioned sequence and False another time. The 'not in' evaluates to true if it does not find a variable in the specified sequence and false otherwise. In the code below, 'A' is a member of the string and hence the print statement outputs True. In the second statement as 's' is not a member of the string and hence the print statement outputs True.

```
a = 'Albert'
print('A' in a ) # Results in True
print('s' not in a) # Results in True
```

Identity Operators

Identity operators are used for checking the memory locations of two objects. The 'is' operator results to True if the variables on either side of the operator point to the same object or False another time. The 'is not' evaluates to False if the variables on either side of the operator point to the same object and True another time. In the code below the first print statement outputs a False and the second print statement outputs a True as in both cases the values are identical (Guzdial and Ericson 2015).

```
a = 10
```

Table 1.6 Python logical operators:

Operator	Command a = 15, b = 0	Result
and (Logical AND)	(a and b)	False
or (Logical OR)	(a or b)	True
not (Logical NOT)	not(a and b)	True

```
b = 10
x = 'Albert'
y = 'Albert'
print(a is not b)     #Results in False
print(x is y)         #Results in True
```

Python Operators Precedence

Table 1.7 gives the precedence of the various operators provided in Python. If the operator is placed higher in the precedence table, it will be executed before the operators at the bottom when they appear in an expression. The operators appearing in the same level/row will have the same precedence. Operators having the same precedence have associativity from left to right (except for exponential operator which has right to left).

Table 1.7 Operator precedence

Operator and description
** Exponentiation (raise to the power)
~, +, − Complement, unary plus, and minus (method names for the last two are + @ and -@)
*, /, %, // Multiply, divide, modulo, and floor division
+, - Addition and subtraction
>>, << Right and left bitwise shift
& Bitwise "AND"
^, | Bitwise exclusive 'OR' and regular 'OR'
<=, <,>, > = Comparison operators
<>, = =, ! = Equality operators
=, % =, / =, // =, - =, + =, * =, ** = Assignment operators
is, is not Identity operators
in, not in Membership operators
not , or, and Logical operators

1.3.5 Python—Decision-Making

In programming, decision-making is done by anticipating a condition to occur in the execution process and then specifying a list of commands to be executed, if that condition occurs.

Various decision structures are employed which evaluate expressions that produce a TRUE or FALSE result. A set of actions have to be performed if the outcome is TRUE and a different set of actions have to be performed if the outcome is FALSE.

Figure 1.6 shows the general control flow in most programming languages when a decision has to be made by the program.

Python has various decision-making statements such as *if...statements, if...else statements and if...elif... else statements.*

An *'if' statement* is made up of a boolean expression followed by single or multiple statements. An *'if...else' statement* has an additional *'else' statement* executed when the boolean expression results in a FALSE value. In an *'if...elif...else' statement* the 'elif' is short for 'else if'. It allows us to check for multiple expressions. If all the conditions are FALSE, the body of else is executed. The below Python program illustrates the use of the above decision-making structure:

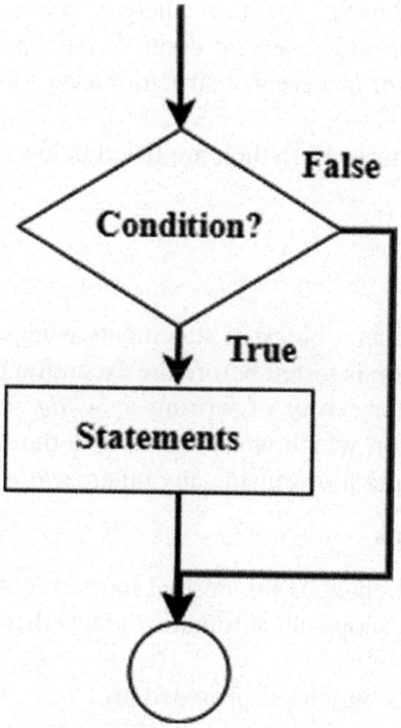

Fig. 1.6 Decision flow in programming languages

```
# if statement in Python
a = 50
if(a==50):
   print("the value is 50")

# if.....else statement in Python
if(a==50):
   print("the value is 50")
else:
   print("the value is not 50")

# if....elif....else statement in Python
a=10
if(a==50):
  print("the value is 50")
elif( a> 50):
  print("the value is greater 50")
else:
  print("the value is less than 50")
```

Loops

In a programming laguage, statements are executed in a sequential manner. A programmer may be faced with a situation where he needs to execute a block of code multiple times. Python provides several control structures which allow the above situation. A loop in Python is a control structure used to run a block of statements multiple times.

The loop control structures in Python are listed below:

- while loop
- for loop
- nested loop

The '*while*' loop repeats a block of statements when a condition evaluates to a TRUE value. The condition is tested before the execution of the block of statements. The '*for*' loop is a concise way of writing a '*while*' loop, as such it is said to abbreviate the code section which is used to update the loop variables. The nested loop structure uses multiple loops inside any other '*while*'or '*for*' loop.

Loop Control Statements

A loop control statement changes the normal loop execution. It must be noted that when execution has left a scope, all automatic objects that were created in that scope are destroyed.

The control statements which are provided are

- break statements
- continue statement
- pass statement

The '*break*' is used to terminate the loop and the program control is transferred to the first statement after the loop. The '*continue*' is used to skip the rest of the body of the loop and to immediately re-evaluate the condition before reiterating. The *'pass'* in Python results in no operation . It may be used by the user to avoid error when empty code is syntactically not allowed in cases such as loops, function definitions, etc.

1.3.6 Functions

A function is defined as a reusable block of related and organized code that can be used to perform a task. A function is principally used to provide modularity. A function called multiple times as such promotes the reusability of code. Python language has a repository of a large number of built-in functions such as print(),input() etc. In addition, Python also allows users to define a function to perform a user-defined task, such functions are called as user-defined functions.

Function Definition in Python

The following set of rules needs to be followed to define a function:

- The keyword *'def'* followed by the function name and parentheses '()' is used to define a function block.
- Input arguments are placed within parentheses.
- An optional first statement in a function is the documentation string of the function or *docstring*.
- Indentation must be followed in a function and the indentation begins after the ":".
- A function has a *'return [expression]'* statement. Its main task is to exit a function or passing back an expression to the caller. We can use return statements without arguments. Below is the example of a function definition:

Syntax

```
#Example
def multiply(a):
    "function to multiply"
    ans = a * a
    return [ans]


val = multiply(50)
print(" the value of multiplication",val)
```

1.3.7 Modules

A module is used to arrange code in order to make it more readable. Grouping similar code into modules makes it easier to comprehend and distribute the code. A module is a Python object containing randomly named characteristics that may be bound and referenced. A module contains function definition, classes, and variables. One can include runnable code in a module.

When a module is created in Python whose name is *test.py*, the corresponding code is written to the file named *test.py*. The file *test.py* contains the following for the definition of a function to find a square of a number.

```
def square(val):
   print("The squre is :",val*val)
   return
```

The *import* Statement

Now the above *test.py* is used as the source file for the module. By executing an import statement in some other Python source file one can import the above source code(*test.py*). The syntax for *import* is as follows:

```
import module_name
```

When the interpreter comes across an import statement, it imports the module if it is on the search path. A search path is a directory list that the interpreter looks over before importing a module. To import the module *test.py*, use the following commands:

```
#!/usr/bin/Python
import test
test.square(15)
```

1.3.8 Files I/O

Printing to the Screen

The print statement is used to display output on the screen. A comma is used to pass more expressions to the output. The print function converts the expression to a string and then writes to the standard output. Whatever is in the double quotes inside a print statement will appear as it is on the output. An example of a print statement is given below:

```
print("Hello to the Python World")
```

Reading Keyboard Input

A built-in function can be used to read a line of text from standard input. The *input([prompt])* function is used to read one line from standard input. The string inside the brackets is displayed as it is on the output, as a prompt message to the user. This function returns the input as a string after the removal of the newline at the end of the input. Whatever is input through the keyboard is displayed on the screen.

```
input_data = input("Please enter a input")
print(" The input by the user is :", input_data)
```

1.4 Python for Embedded System

According to IEEE Spectrum, presently Python is positioned No. 1 among the programming languages. Python has carried on its skyward ambit since 2016 and jumped to No. 1 in the overall rating in the year 2017. This is a very interesting programming language, but it still requires to make an impact in the embedded field. There are *Zerynth* and *MicroPython* platforms, which are on rise for embedded system development. As we are aware, the embedded code is still written in C. C programming experts justify this stating that "*C has faster runtime and is more compact*".

… BUT…

if one replaces C with "assembly" in the above declaration, this is exactly what the programmers said about 20 years ago!

Can Python be Used for Embedded Systems? Over the years, the C/C++ programming languages have dominated the embedded systems programming. On the other side, Python has many strengths that make it a great language for embedded programming.

Python isn't only the most popular language, it's also the fastest growing language for embedded computing. May be that sounds silly, but the reality is that Python has started eating into C/C++'s space.

Python can be used in embedded, small, or minimal hardware devices, depending on how limiting the devices actually are.

The question arises: **Can Python be used to program microcontrollers?**

MicroPython is an efficient and optimized implementation of the Python 3 programming language that allows the programming of microcontrollers. MicroPython is a small subset of the Python standard library and is optimized to run on microcontrollers and in constrained environments.

We will also talk about PyBoard, a microcontroller board specifically designed to be programmed using MicroPython.

1.5 Introduction to IoT

The term IoT, i.e. "Internet of Things" (IoT) was initially coined by Kevin Ashton in 1999, which later became widely accepted throughout the world. IoT is a mesh of physical objects or people called "things" that are encapsulated with software, hardware, network connectivity, and sensors, which admits these objects to gather and barter the data. IoT makes everything "smart" and "easy" by improving aspects facet of individuals' life with the potential of data collection, Artificial Intelligence algorithm, and connectivity. There are basically four basic components of an IoT system, namely, sensors, connectivity, data processing, and user interface as shown in Fig. 1.7

Sensors: Sensors/devices are important parts of IoT system which help to gather live data from the surrounding. This data is complex and varied in nature. Sometimes a device may have other types of sensors which perform some kind of intelligent task apart from sensing.

Connectivity: Connectivity is another component of IoT. All the data gathered from the sensors is sent to a cloud. These sensor data will be sent to the cloud by using several communication protocols such as Bluetooth, wireless, NFC, RFID, radio protocols, and Wi-Fi-Direct.

Data Processing: Once the data is gathered and uploaded on to the cloud, the various available data processing software performs required processing algorithm.

User Interface: After performing the processing on data, the processed data must be made available to the end user in some way. The end user gets these data on his smart device. Sometimes user from distant place can control the various devices through IoT.

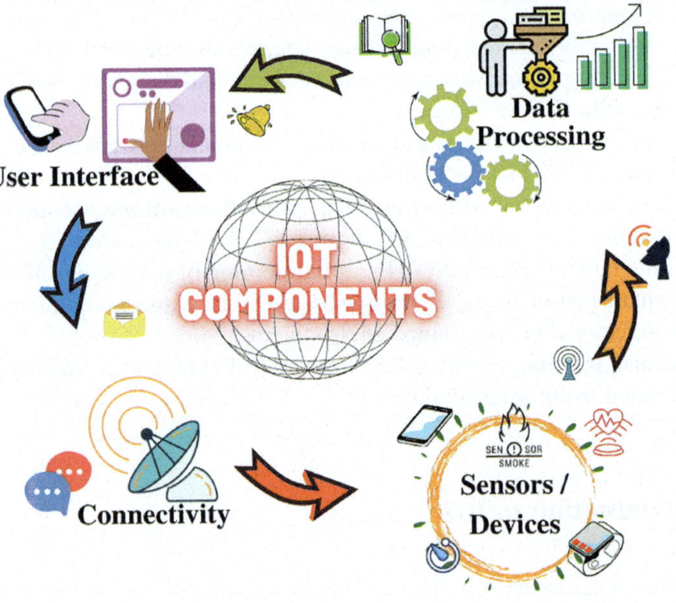

Fig. 1.7 Basic components of IoT

1.6 IoT Applications

IoT helps to create opportunities to connect and interact with physical world by using sensors and Internet. IoT system allows devices to be controlled remotely over the Internet. The interconnection of these various devices helps in automizing the entire system. This results in enhanced accuracy, system efficiency, and economic benefit with minimal human intervention.

IoT devices find extensive set of applications into consumer, commercial, industrial, and infrastructure spaces (Vongsingthong and Smanchat 2014; Jump up to 2015, Perera et al. 2015). Below are the list of few applications of IoT.

Energy Applications: IoT find its extensive use in energy sector. The energy saving is very important, so one has to utilize it very effectively. Individuals and organizations are finding out various ways to minimize and control the energy consumption. With advancement in IoT technologies it makes it possible to monitor the energy consumption at the device level, individuals house level, electric grid level, and also at distribution level. It also helps to monitor the system performance.

Smart Homes/Smart City: This is one of the most frequently used practical applications of IoT. Smart homes system ensures both convenience and home security. Under smart city, it allows Internet access to people. IoT helps to manage traffic, waste management, water distribution, electricity management, etc. This will help to give some kind of convenience to the city management administration.

Healthcare Application: In today's world health monitoring is very important. In present state, people need to monitor their own health at regular intervals. Smart devices on patients continuously updates the data on cloud and notifies this to the hospital/doctor in case of emergency. So, if a patient is coming to the hospital in case of emergency, by the time he or she reaches the hospital his health report is diagnosed by doctors and the hospital can quickly take the treatment decision which is very much crucial.

Education: IoT plays an important role in education. IoT in education system means better connectivity for learning which helps to fulfill the gaps in the education sector. Today, many schools and colleges have IoT embedded into their day-to-day learning. It improves the overall quality of education.

Air and Water Pollution: IoT also plays an important role in monitoring the environmental pollution. There are various sensors available which help to detect pollution in the air and water. By employing IoT system, one can automize and reduce the human intervention in monitoring.

Agriculture: Agriculture is the main stay of our economy. Smart agriculture is very important these days as farmers will have ready information about various agricultural parameters. IoT device installed at remote places measures various parameters such as soil moisture, chemical application, dam levels, and provides this information to the farmer in real time.

Transportation: In transportation sector, IoT has changed the entire facet of the same. Now, the market has seen introduction of automated cars with embedded sensors, which can sense the traffic and switch off automatically. IoT also helps to identify the location of free parking slot. Some cars are embedded with sensors which can monitor the current health status of the vehicle, so that one doesn't face any issues while traveling.

Marketing your product: IoT has ventured into the marketing industry, using IoT, companies can now analyze and respond to customer preferences by providing the desired content and solutions. IoT helps enhance the business strategies in the real time.

Conclusion:

Python programming and IoT chapter gives the brief introduction to Python and history of various Python versions. As we all know, compared to other programming languages, C/C++ is preferred by programmers for embedded applications. But in recent past Python is extensively used for embedded application and it is eating the C/C++ space. Developers shouldn't be surprised when they see Python cropping up and beginning to play a vital role in embedded system development. An overview of the Python programming with detailed installation steps of various Python IDEs such as Thonny, Pycharm, and Anaconda is also provided. There's a lot of noise at the moment about the IoT and its impact on everything. The coding flexibility and dynamic nature of Python helps developers in creating intelligent IoT devices. Introduction to IoT, its working, and few examples are covered nicely in this chapter.

References

Beazley D, Jones BK (2013) Python cookbook: recipes for mastering python, vol 3. O'Reilly Media, Inc.

Guzdial MJ, Ericson B (2015) Introduction to computing and programming in Python. Pearson

Jump up to: The enterprise internet of things market. Business Insider. 25 February 2015. Accessed 26 June 2015

Kuhlman D (2012) A python book: beginning python, advanced python, and python exercises. Section 1.1. Archived from the original (PDF) on 23 June 2012

Perera C, Liu CH, Jayawardena S (2015) The emerging internet of things marketplace from an industrial perspective: a survey. IEEE Trans Emerg Top Comput 3(4):585–598. arXiv:1502.00134. Bibcode:2015arXiv150200134P. https://doi.org/10.1109/TETC.2015.2390034. ISSNS 2168-6750. S2CID 7329149

Pilgrim M, Willison S (2009) Dive into Python 3, vol 2. Springer

Vongsingthong S, Smanchat S (2014) Internet of things: a review of applications & technologies (PDF). Suranaree J Sci Technol

Chapter 2
Configuring Raspberry Pi, MicroPython Pyboard, and Jetson Nano for Python

Abstract In today's world one need not spend much time in configuring and getting acquainted with any embedded platform. The main focus should be towards quick launch of any designed application to market. Keeping these points in mind authours have provided the detailed steps involved in configuration and programming the platform. Here, authours have focused on three different embedded Platforms such as Raspberry pi Board for basic interface application, MicroPython Py board for IoT applications and NVIDIA Jetson Nano board for Machine learning and deep learning Applications using Python programming with FoG/cloud Computing.

Keywords Raspberry pi · MicroPython Py · Jetson Nano · Firmware · Network interface card · Single board computer

2.1 Raspberry Pi Board Features

Raspberry Pi is a small single-board computer (SBC). By connecting peripherals like keyboard, mouse, display to the Raspberry Pi, it will act as a mini personal computer. Raspberry Pi is popularly used for image/video processing, IoT, and also robotics applications. Raspberry Pi will have speed limitations as compared to a general-purpose desktop but can still provide comparable capabilities with much lower power consumption and cost. Debian-based NOOBS Operating Systems (OS) and Raspbian OS are the official operating systems for Raspberry Pi. In addition, there are other OSs such as Archlinux, Ubuntu, Windows 10, RISC OS, IoT Core, etc. which are adapted for the Raspberry Pi platform. Currently, the Raspbian OS is optimized for use on the Raspberry Pi. Task such as Python programming, browsing, word processing, games, can be efficiently handled on the Raspbian OS. The microSD card with an 8 GB minimum storage is recommended for the smooth functioning of the OS. Here we have used Raspberry Pi 3 B+ boards as shown in Fig. 2.1.

The original version of this chapter was revised. Series editor name has been removed from online version of this chapter. The correction to this chapter is available at https://doi.org/10.1007/978-981-19-9466-1_7

J. S. Parab et al., *Python Programming Recipes for IoT Applications*, Transactions on Computer Systems and Networks, https://doi.org/10.1007/978-981-19-9466-1_2

Fig. 2.1 Raspberry Pi 3B+ board

General-Purpose Input Output (GPIO) is provided on the Raspberry Pi which can be used for a myriad of applications such as control of motors, sensors, etc. The computing unit of the Raspberry Pi is the ARM-based Broadcom Processor SoC which has an on-chip GPU(Graphics Processing Unit). The clock speed offered for the processor range from 0.7 to 1.2 GHz based on the various models of the Raspberry Pi. SDRAM is provided with capacities ranging from 256 MB to 1 GB. Popular serial interfaces such as the SPI, I2C, I2S, and UART are provided on the Raspberry Pi. The different versions of the Raspberry Pi are Raspberry Pi Zero, Raspberry Pi 1 A, Raspberry Pi 1 A+, Raspberry Pi 1 B, Raspberry Pi 1 B+, Raspberry Pi 2 B, Raspberry Pi 3 B, Raspberry Pi 3 B+, and Raspberry Pi 4.

Out of the above versions of Raspberry Pi, more prominently used Raspberry Pi models and their features are given in Table 2.1.

Table 2.1 Features of different versions of Raspberry Pi

Features	Raspberry Pi model B	Raspberry Pi 2 model B	Raspberry Pi 3 model B+	Raspberry Pi zero
SOC	Broadcom 2835	Broadcom 2836	Broadcom 2837B0	Broadcom 2835
CPU	ARM-11	A7 Quad-Cortex	A53 Quad-Cortex	ARM-11
Operating frequency (GHz)	0.7	0.9	1.4	1
RAM (SDRAM)	512 MB	1 GB	1 GB	512 MB
GPU speed (Videocore IV)	0.25 GHz	0.25 GHz	0.4 GHz	0.25 GHz
Storage (microSD)	Supported	Supported	Supported	Supported
Ethernet	Supported	Supported	Supported	Not supported
Wireless	Not supported	Not supported	Bluetooth and Wi-Fi	Not supported

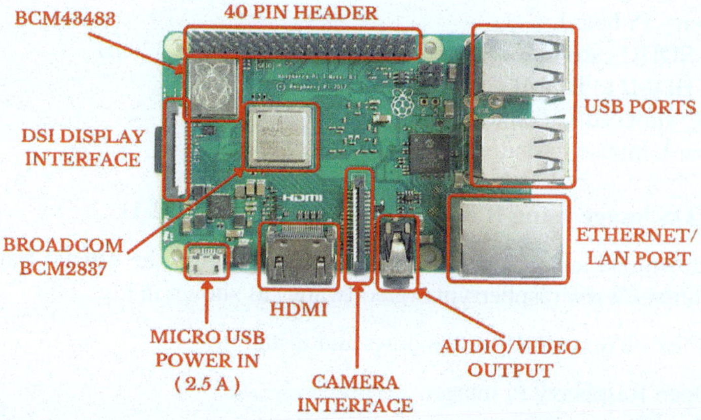

Fig. 2.2 Raspberry Pi 3 model B+ hardware

Raspberry Pi 3 B+ Hardware

The hardware on the Raspberry Pi 3 B+ is highlighted in Fig. 2.2. Important hardware features are explained here. "Raspberry Pi 3 ModelB+". https://static.raspberrypi.org/files/product-briefs/Raspberry-Pi-Model-Bplus-Product-Brief.pdf

- **CSI Camera Interface**: CSI (Camera Serial Interface) connector is used to connect a Pi camera to the Broadcom Processor.
- **HDMI (High-Definition Multimedia Interface)**: This interface is used to transmit video and digital audio in an uncompressed form to a display monitor which supports HDMI.
- **Composite Audio and Video Output**: The port is used to send audio along with the video to the audio/video systems.
- **Power LED**: The red-colored LED gives the power indication. This LED begins to blink when the supply voltage falls below 4.63 V.
- **DSI (Display Serial Interface)**: DSI interface uses a 15-pin ribbon cable to connect to an LCD. This is a fast high-resolution display interface that sends video data directly to the display from the GPU.
- **Activity LED**: This green-colored LED indicates microSD card activity.

2.1.1 Configuration of Raspberry Pi

Flashing OS image on Raspberry Pi

Raspberry Pi OS is open-source based on Debian 10. Debian is a popular open-source Linux distribution. Two methods are generally used to install the Raspberry Pi OS on a blank microSD card. Before starting, make sure you have the following prerequisites:

- Raspberry Pi board.
- Micro-SDHC card (with SD adapter) of size at least 8–32 GB.
- Micro-HDMI to HDMI cable.
- 5 V DC via USB-C connector.
- Keyboard, mouse, monitor.

Flashing OS image using Raspberry Pi Imager (Method 1)

Step 1: Download and install Raspberry Pi Imager from the official Raspberry Pi website (https://www.raspberrypi.org/software/) as shown in Fig. 2.3a.

Step 2: Click on download for windows and install it.

Step 3: Open Raspberry Pi imager.

Step 4: Choose OS as shown in Fig. 2.3b.

Step 5: Select 'Raspberry Pi OS' as shown in Fig. 2.3c.

Step 6: Insert microSD card and it will get detect automatically as shown in Fig. 2.3d and e.

Step 7: Click on 'Write' as shown in Fig. 2.3f.

Flashing OS image using balenaEtcher (Method 2)

Step 1: Download Raspberry Pi Desktop OS from https://www.raspberrypi.org/software/ by clicking on the Download Raspberry Pi Desktop button as shown in Fig. 2.4a.

Step 2: Click on the 'Download' button as shown in Fig. 2.4b https://www.raspberrypi.org/software/raspberry-pi-desktop/).

Step 3: Download balenaEtcher from https://www.balena.io/etcher/ and install it. Etcher enables the user to create bootable USB flash drives.

Step 4: There are three versions of Raspberry Pi OS, namely, Raspberry Pi OS (32 bit) Lite, Raspberry Pi OS (32 bit), and Raspberry Pi OS Full (32 bit) are available.

Step 5: While using balenaEtcher, select the option 'Flash from file' and browse to the location where the image is downloaded.

Step 6: The microSD card needs to be formatted. Choose the 'select target' option and select the target as your microSD.

Step 7: The final step is to hit the 'Flash' button and then the balenaEtcher will start the flashing process as shown in Fig. 2.5.

Booting Raspberry Pi for the First Time

Step 1: Insert your microSD card into Raspberry Pi.

Step 2: Connect power, keyboard, mouse, and monitor to Raspberry Pi.

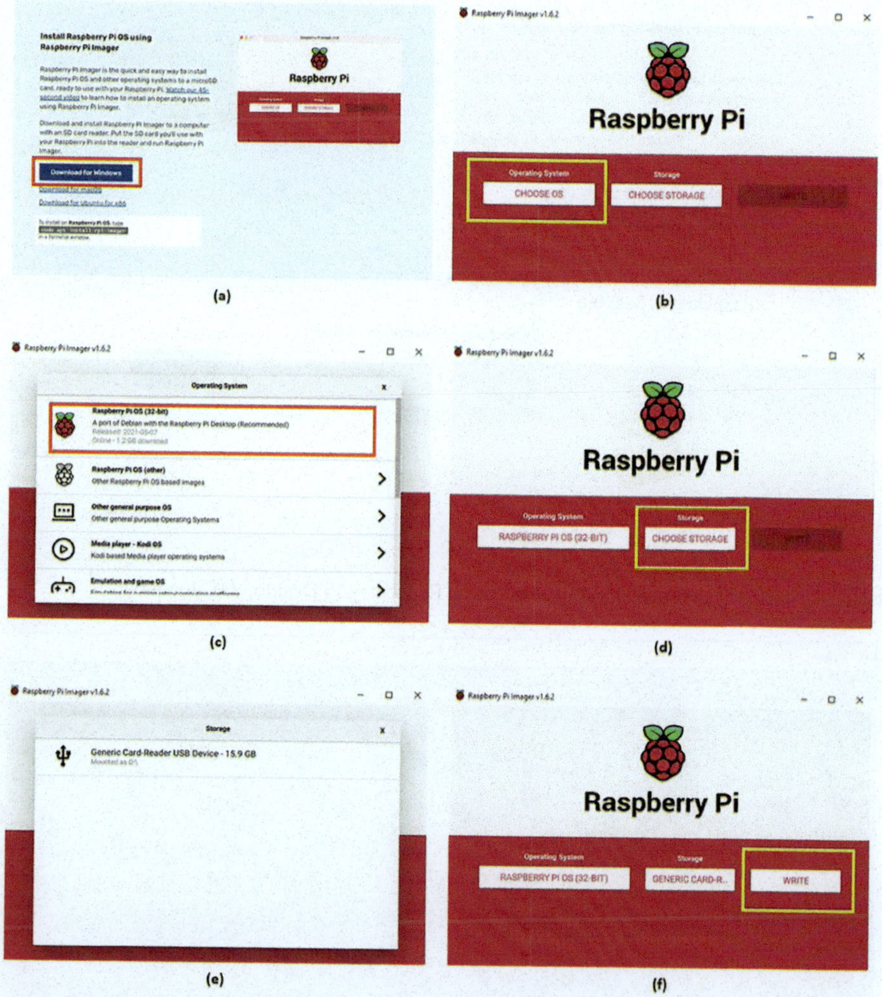

Fig. 2.3 **a** Raspberry website, **b** Raspberry imager, **c** Raspberry OS selection, **d** choose storage for OS installation, **e** microSD card or storage selection, and **f** OS installation

Step 3: After booting you get “Welcome to the Raspberry Pi” dialog on the display, as shown in Fig. 2.6.

Step 4: Quick setup of Raspberry Pi:

(i) On the next screen you will be asked to choose your Country (Fig. 2.7a), Language, and Timezone.
(ii) Set account password as shown in Fig. 2.7b.
(iii) Followed by this, select a Wireless Network and its network password.
(iv) Set up the desktop screen by clicking ’Next’ as shown in Fig. 2.7c.

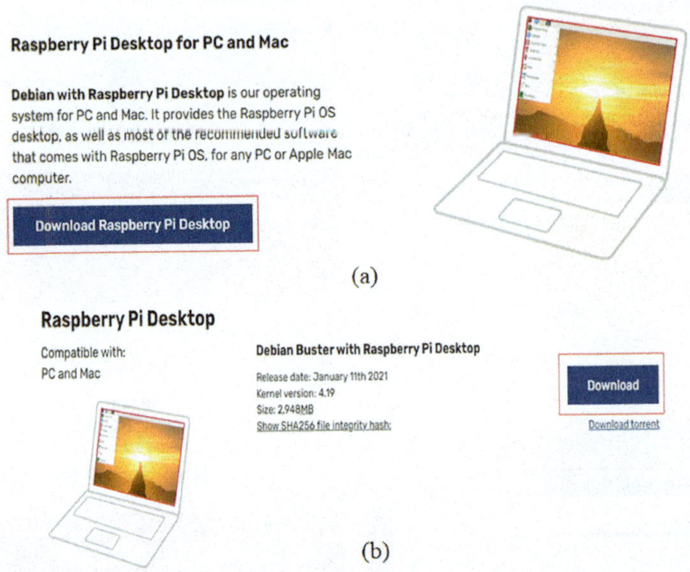

Fig. 2.4 **a** Raspberry Pi Desktop website and **b** Raspberry Pi Desktop OS download

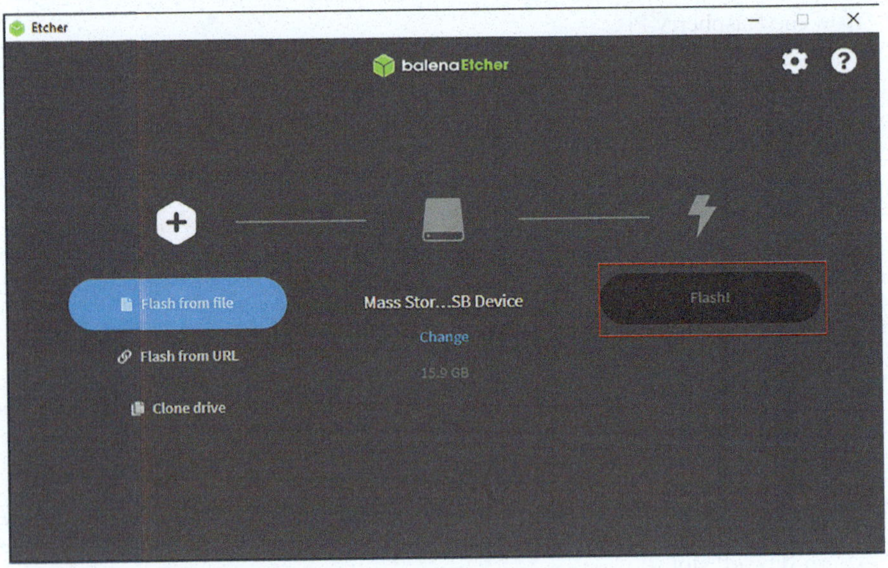

Fig. 2.5 BalenaEtcher for flashing Raspberry Pi

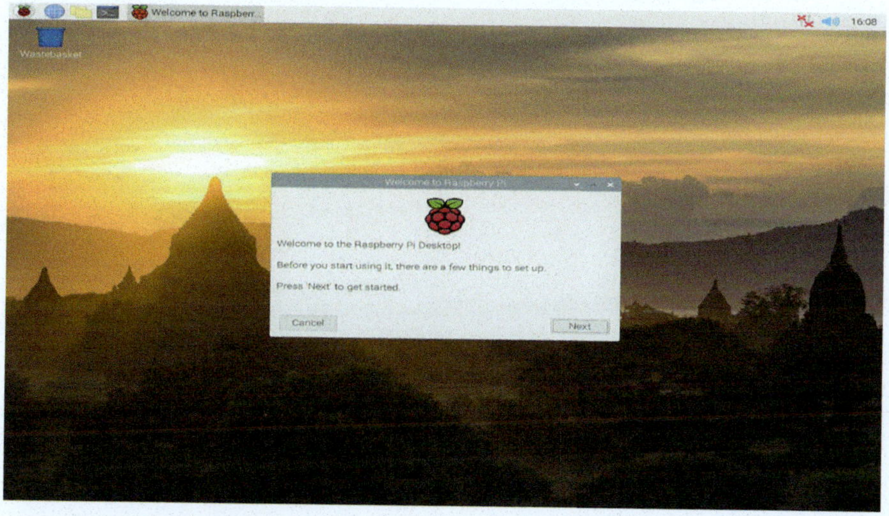

Fig. 2.6 Screen after the first boot

(v) Finally it shows setup complete as shown in Fig. 2.7d.

Now the Raspberry Pi is ready to use.

2.2 MicroPython Pyboard Features

The MicroPython Pyboard is a small form factor development board designed to run MicroPython to realize various IoT projects. MicroPython is versatile and provides an interactive prompt, locks, generators, list understanding, exception handling, etc. Below is the specification of Pyboard V1.1: "Quick reference for the Pyboard". https://docs.micropython.org/en/latest/Pyboard/quickref.html

Pyboard V1.1 Specifications:

- STM32F405RG microcontroller (clock frequency 168 MHz Cortex-M4 core)
- 1024 KB flash ROM, 192 KB RAM
- Power from microUSB connector
- MMA7660 3-axis accelerometer
- MicroSD card slot
- 5 GPIO on the bottom, 24 GPIO on the left and right, Real-Time Clock (RTC)
- Three 12-bit analog-to-digital converters (ADC)
- Two 12-bit digital-to-analog (DAC) converters
- One user and one reset switch, four LEDs
- Wide input voltage range from 3.6 V to 16 V, 3.3 V low dropout voltage regulator
- Upgrade firmware through DFU bootloader in ROM

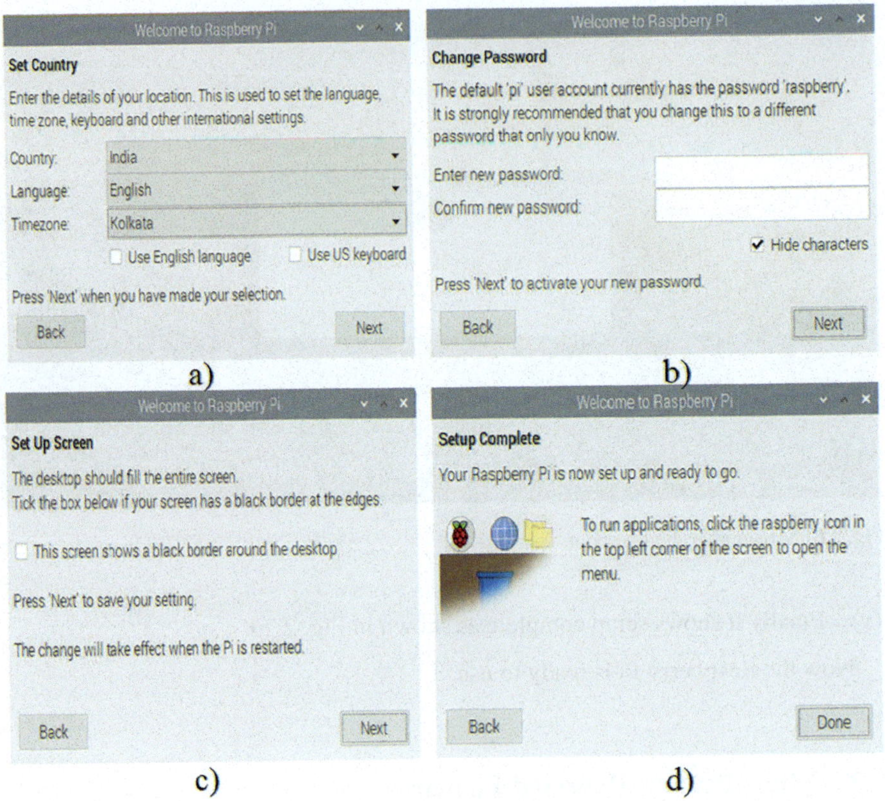

Fig. 2.7 **a** Set country, language, and timezone; **b** set a new password; **c** set up screen for desktop; and **d** setup complete

2.2.1 *Configuration of MicroPython Pyboard*

Loading the MicroPython firmware on Pyboard v1.1

The Pyboard comes with a strong microcontroller of ARM Cortex-M4, as shown in Fig. 2.8. The coding can be started without any installation because the MicroPython interpreter is pre-loaded with the Pyboard. The pin details are given in Fig. 2.9.

The basic steps to start the Pyboard are as follows:

- Connect the Pyboard to a laptop or computer and list it as a USB storage device.
- Navigate to the Pyboard drive (likely called "PYBFLASH").
- Edit '*main.py*' with MicroPython code and save the file.
- Select the drive from the OS.
- To start executing code or application press the Reset button (RST).

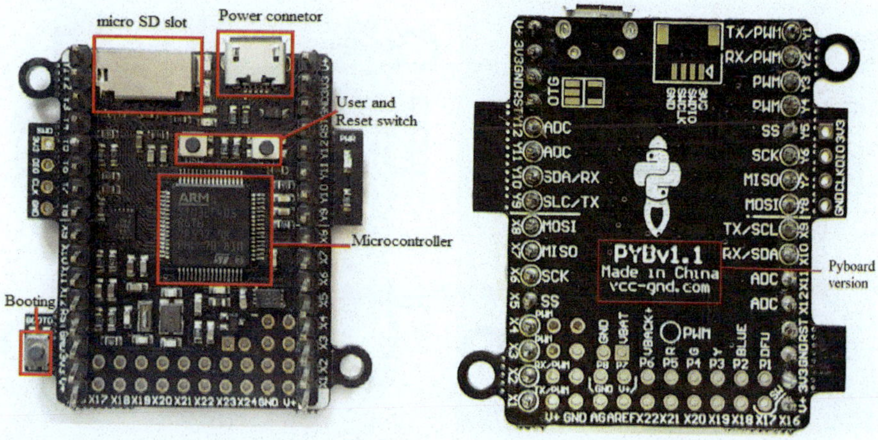

Fig. 2.8 Front and back of Pyboard v1.1

Fig. 2.9 Pyboard pin details

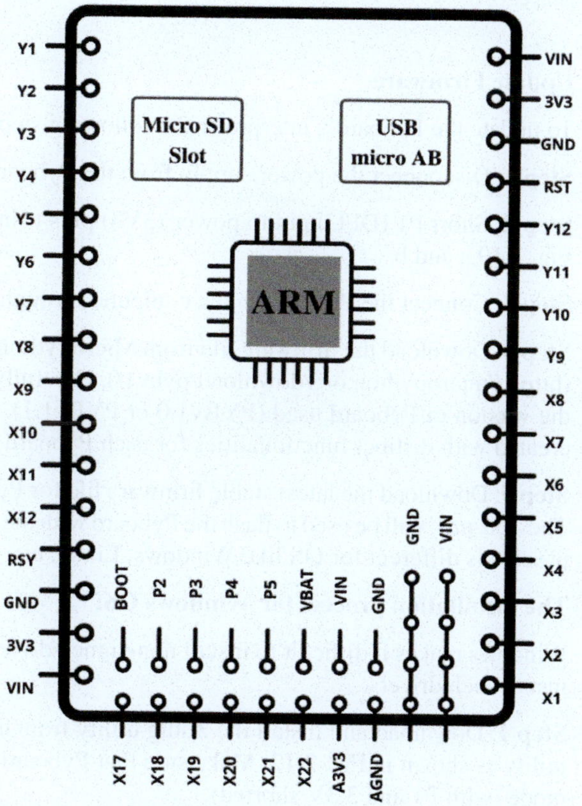

(a) (b)

Fig. 2.10 **a** Connect P1 (DFU) and 3V3 pins and **b** connection of P1 (DFU) and 3V3 pins using jumper

Update Firmware

To update the Pyboard's interpreter the following steps need to be followed:

Step 1: Disconnect the power supply from the Pyboard.

Step 2: Short P1 (DFU) pin to power (3V3) pins using jumper or wire as shown in Fig. 2.10 a and b.

Step 3: Connect the Pyboard to the computer through the USB port.

Step 4: Download the firmware file from MicroPython website as shown in Fig. 2.11 (https://micropython.org/download/pybv1/). Carefully select the firmware based on the version of Pyboard used (PYBv1.0 or PYBv1.1). Several firmware versions are created with distinct functionalities for each Pyboard version.

Step 5: Download the latest stable firmware file for Pyboard having extension '*.dfu*'. The '*dfu-util*' will be used to flash the Pyboard with the new firmware. The installation process is different for OS like Windows, Linux, etc.

The installation process for Windows OS:

Windows makes it difficult to install non-signed drivers, so use a special program to install such drivers.

Step 1: Download and install the Zadig utility from https://zadig.akeo.ie/. Open the utility as shown in Fig. 2.12. Make sure that Pyboard is plugged in and it is in DFU mode (with P1 and 3.3V shorted).

Step 2: In the Zadig utility, select Options → 'List All Devices'. From the drop-down menu, select 'STM32 BOOTLOADER'. Zadig utility recommends replacing the 'STTub30' driver with 'WinUSB', as shown in Fig. 2.13. If you are unable to see the installed driver then select WinUSB. Successful installation is shown in Fig. 2.13.

Step 3: From http://dfu-util.sourceforge.net/releases/dfu-util-0.9-win64.zip download the latest release of *dfu-util* firmware. Unzip the folder and open a command prompt, as shown in Fig. 2.14. Navigate to the unzipped folder. If the unzipped file is saved in the Downloads directory, then use the following command (Fig. 2.14) or give the correct path of saved files:

cd Downloads\dfu-util-0.9-win64\dfu-util-0.9-win64.

To upload the latest firmware to the Pyboard (Fig. 2.11), run the following command:

```
dfu-util --alt 0 -D
```

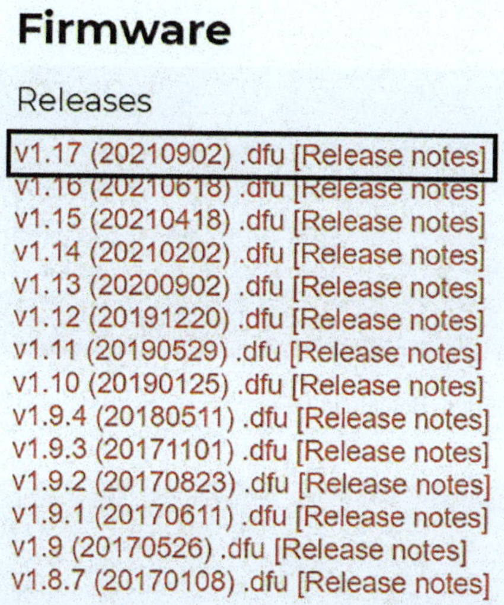

Fig. 2.11 Latest MicroPython firmware

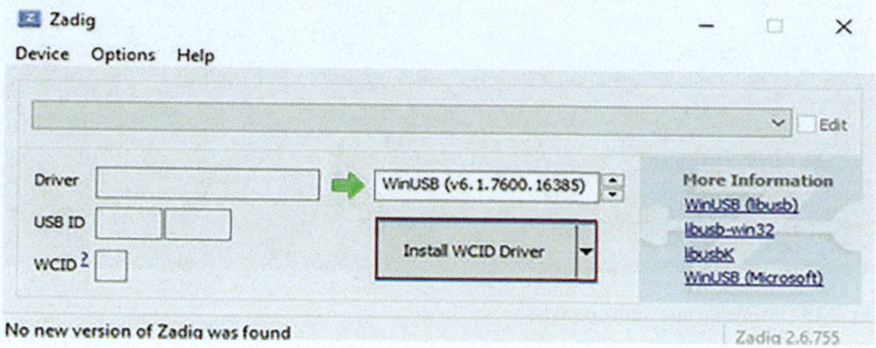

Fig. 2.12 Zadig utility

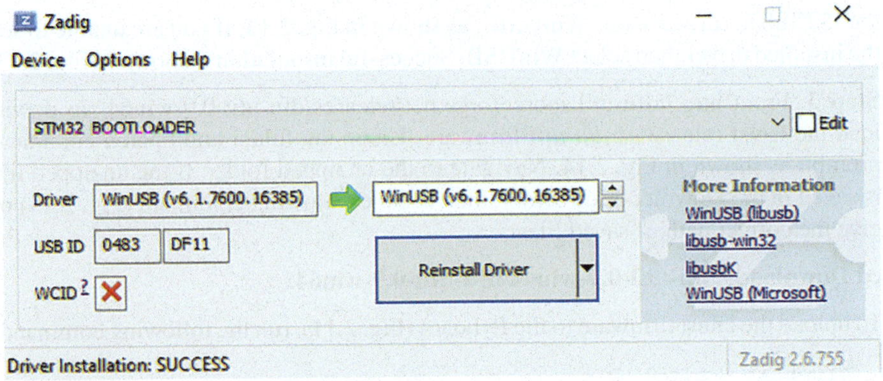

Fig. 2.13 After sucessful installation

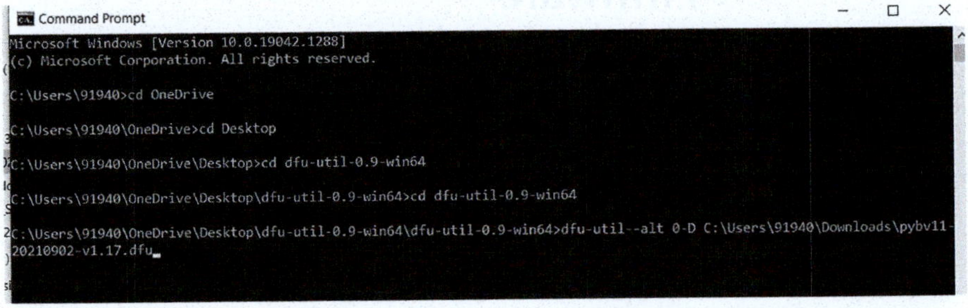

Fig. 2.14 Command prompt to download DFU

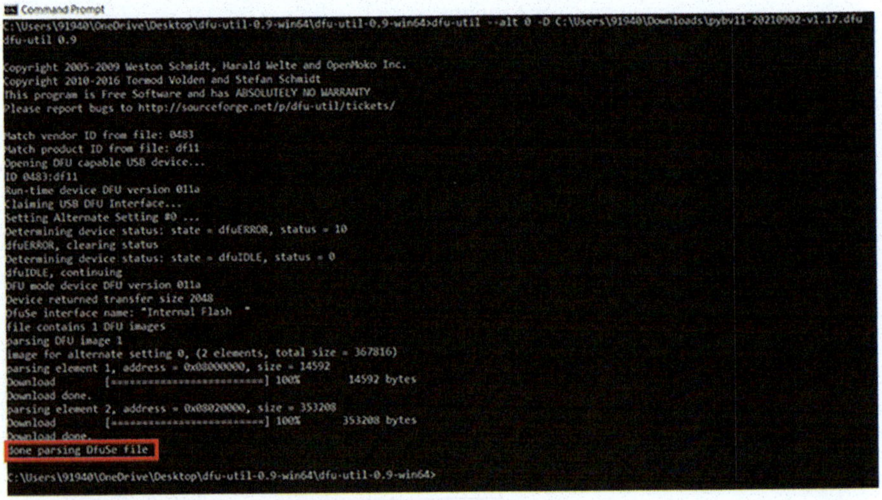

Fig. 2.15 Firmware uploading process

```
<path to .dfu>
<path to .dfu>
```

is the complete path of the downloaded Pyboard firmware, as shown in Fig. 2.14

Step 4: If the firmware was successfully uploaded, you should get a message that ends with *'done parsing DfuSe file'* as shown in Fig. 2.15.

Step 5: To test Pyboard, connect it to USB port of a machine. Open the *'Device Manager'*, and identify the' *Port (COM and LPT)'*. If it shows 'USB Serial Device', then the driver is installed and working properly, as shown in Fig. 2.16. Manually install the driver if you see a yellow exclamation mark(!). Right-click on the *'device'* → choose *'Properties'* → choose *'Driver Install'*. Select the 'PYFLASH' drive (Fig.2.17) to launch the file explorer. If you have not removed the Pyboard flash memory, the '.inf' driver should be saved in the flash. Click *'OK'* → *'Next'*, and the driver should be installed. The *USB Serial Device* listed under *Ports (COM & LPT)*. Write down or copy the COM name (e.g. *COM6*), as shown in Fig. 2.16.

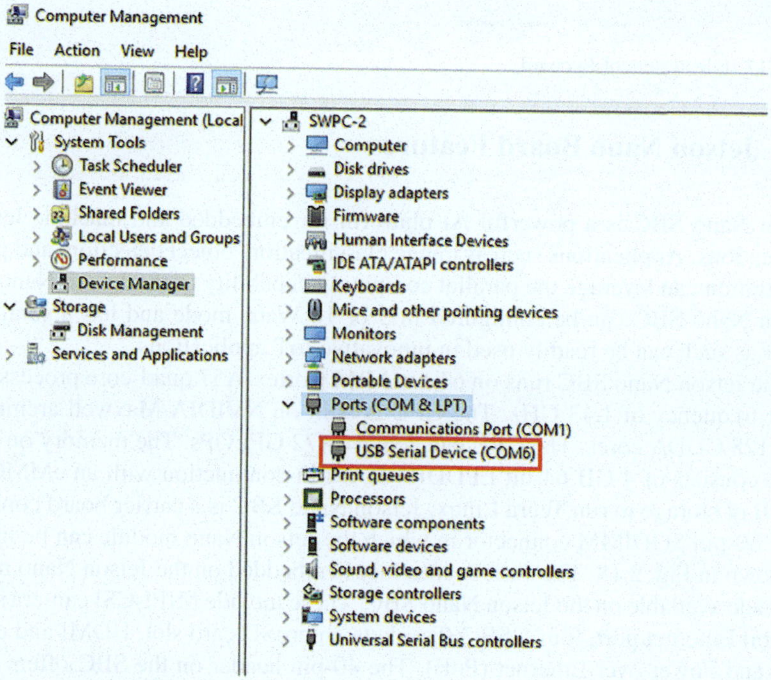

Fig. 2.16 Device manager entry of Pyboard

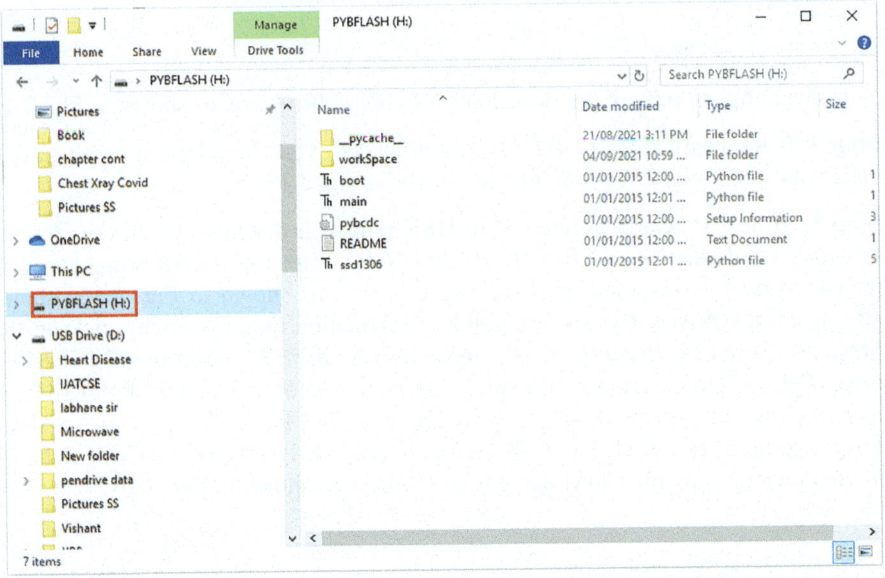

Fig. 2.17 File system of Pyboard

2.3 Jetson Nano Board Features

Jetson Nano SBC is a powerful AI platform for embedded and machine learning applications. Applications such as image classification, object detection, and pattern recognition can leverage the parallel computing capability of the Jetson Nano SBC. Jetson Nano SBC can be configured in 5 or 10 Watts mode and has a small form factor as such can be readily used in innovative IoT applications.

The Jetson Nano SBC runs on 64-bit ARM Cortex-A57 quad-core processor at a clock frequency of 1.43 GHz. The GPU is built on NVIDIA Maxwell architecture with 128 CUDA cores. The GPU can achieve 472 GFLOPs. The memory on Jetson Nano consists of 4 GB 64-bit LPDDR4 RAM in conjunction with an eMMC with 16 GB of storage to run Tegra Linux. Jetson Nano SBC is a carrier board consisting of a 260-pin SODIMM connector in which the Jetson Nano module can be inserted as shown in Fig. 2.18. The various interfaces embedded on the Jetson Nano module are made available on the Jetson Nano SBC. These include MIPI-CSI camera socket, Gigabit Ethernet port, four USB 3.0 sockets, microSD card slot, HDMI and display port, and Power-over-Ethernet (PoE). The 40-pin header on the SBC offers GPIO, I2C, I2S, SPI, PWM, and UART. The specification of the Jetson Nano Developer kit is listed below (NVIDIA 2019):

- ARM Cortex-A57 quad-core processor
- 128 CUDA core GPU based on NVIDIA Maxwell architecture
- 4 GB 64-bit LPDDR4 at 1600 MHz and 25.6 GB/s

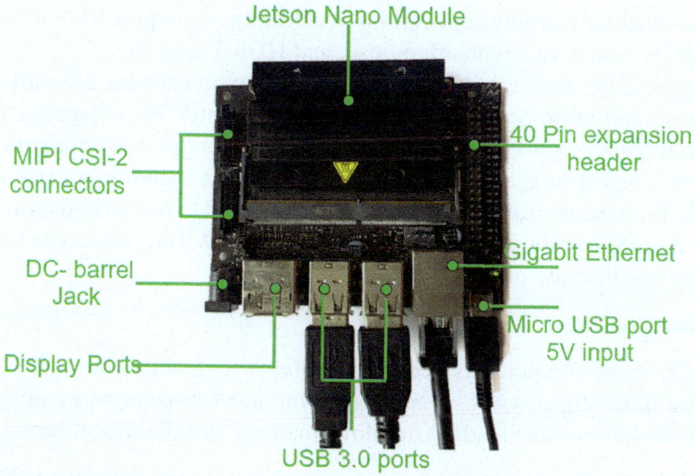

Fig. 2.18 Jetson Nano SBC

- 16 GB eMMC 5.1
- 12 lanes MIPI-CSI-2 D-PHY 1.1
- Gigabit Ethernet
- eDP 1.4 and HDMI 2.0
- M.2 Key E Expansion slot for Wi-Fi
- Interface options—GPIO, I^2C, I^2S, SPI, UART
- MicroSD Storage

The Jetson Nano Developer Kit can be powered by a micro USB 5 V supply with a current rating of greater than 3A. The supply must be greater than 4.75 V to eliminate the brownout condition. There is a DC barrel Jack that can be used to supply a 5 V input and must have a current rating greater than 4A. Make sure that a jumper is put on the J48 header before you use the DC barrel jack for giving the supply voltage.

The Jetson Nano is by no means a stripped-down version of other NVIDIA offering such as Jetson TX2, Jetson Xavier, and Jetson AGX. It is designed to leverage the acceleration offered by the deep learning, graphics, and computer vision libraries included in the JetPack SDK and Deepstream SDK.

2.3.1 *Configuration of Jetson Nano Board*

Flashing the OS Image onto the MicroSD Card

For the flashing process, we require a microSD card with a minimum storage capacity of 32 GB (UHS-1 category). It is recommended to use a card with a capacity of 64 GB or more for machine learning applications. In addition, we require a micro USB power supply (5VDC and 3A), keyboard, mouse, and HDMI display.

We require a PC with a built-in SD card reader or an external SD card adapter to perform read and write operations on the microSD card. The image for the Jetson Nano Developer Kit can be downloaded from the link: (https://developer.nvidia.com/jetson-nano-sd-card-image). Make sure you have a free memory of more than 6 GB on your PC for a successful download. The steps needed to format and flash the image on to the microSD card will vary based on the OS used. The steps to be followed on a Windows machine are given below.

Formatting the MicroSD Card

Step 1: SD card formatter is required initially to format the card. Download it using the link: https://www.sdcard.org/downloads/formatter/sd-memory-card-formatter-for-windows-download/. After downloading install the software.

Step 2: Select the drive associated with your microSD card as shown in Fig. 2.19.

Step 3: Select the quick format option under formatting options. Leave the volume label field blank.

Step 4: Finally click the 'format' button to begin the process.

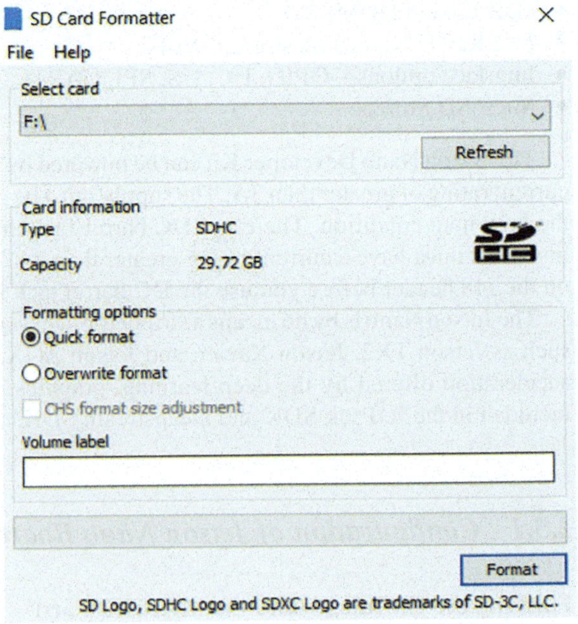

Fig. 2.19 SD card formatter

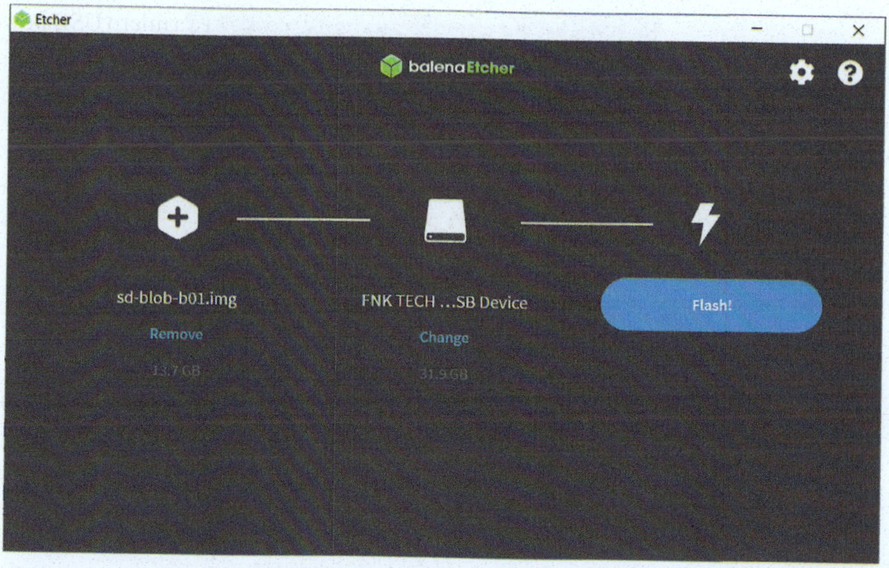

Fig. 2.20 Flashing the OS image on Jetson Nano

Flashing the OS Image

BalenaEtcher software is used to flash the OS image to the microSD card on the Jetson Nano SBC. It can be downloaded using the link: (https://www.balena.io/etcher). The same steps followed for flashing OS image of Raspberry Pi needs to be followed here with the appropriate OS image. The balenaEtcher user interface before flashing is shown in Fig. 2.20. Connect the display, keyboard, and mouse to begin with the setup and booting.

The First Boot of Jetson Nano

The microSD containing the OS image card must be inserted in the slot on the underside of the SODIMM Jetson Nano module. Figure 2.21 shows the position of the slot and inserted microSD card.

Power up the Jetson Nano SBC using the micro USB connector. The barrel jack can be used as the alternate means to supply +5V DC input. The monitor must be connected via the HDMI port. Also, connect the USB mouse and the USB keyboard. Next, the green LED near the micro USB will turn ON and the Jetson Nano SBC will boot for the first time as shown in Fig. 2.22.

The first boot for the SBC will take you through some initial setup screens which include the review and acceptance of NVIDIA Jetson software end-user license agreement screen. Next is the language, time zone, and keyboard layout selection screen. Computer name, username, and password input screen follow as shown in Fig. 2.23. This is followed by the APP partition size selection screen.

Fig. 2.21 The slot for microSD card

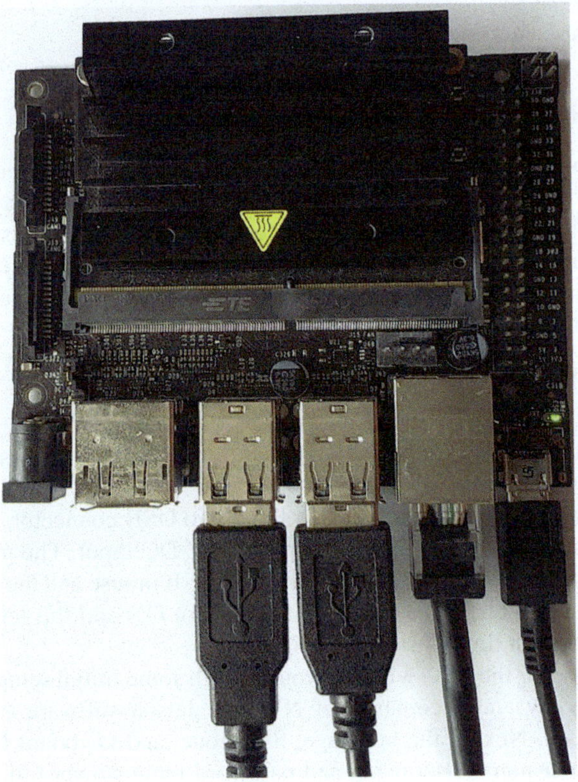

Fig. 2.22 First boot of the Jetson Nano

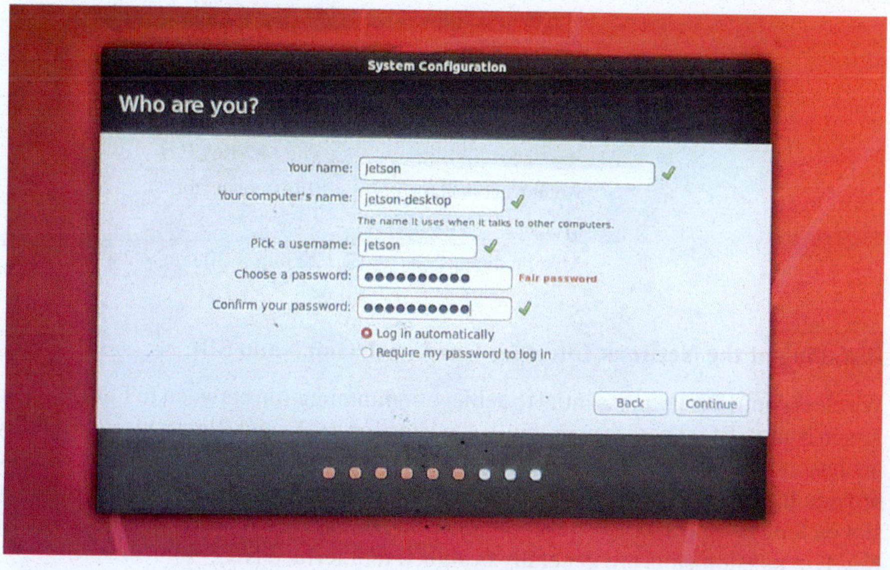

Fig. 2.23 User name, password

Once the Jetson Nano is successfully booted for the first time, you will see a screen with all the keyboard shortcuts as shown in Fig. 2.24.

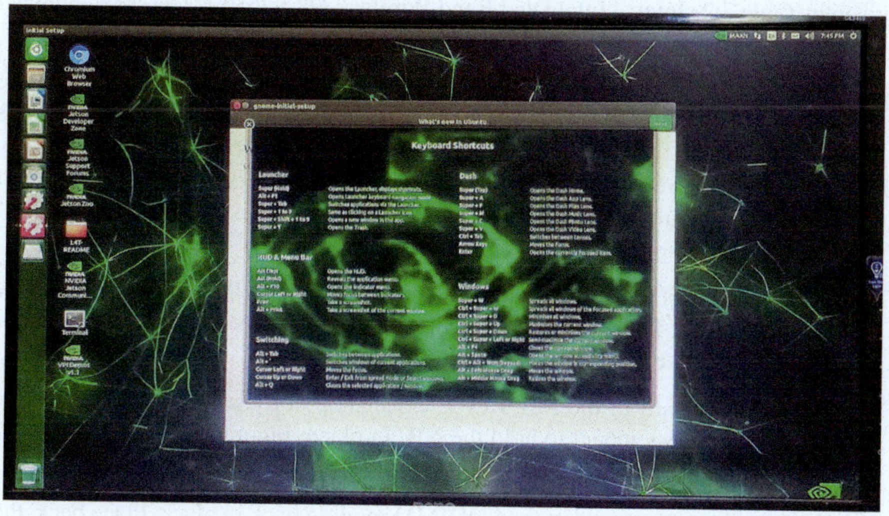

Fig. 2.24 Desktop screen after first boot

Table 2.2 Specification of Intel 8265 AC NIC

NIC chip	Intel 8265AC
Protocol	802.11ac
Operating bands	2.4/5 GHz
Max speeds	300 Mbps/867 Mbps
Bluetooth version	4.2
Interface	The M.2 Key E
Antenna connector	IPEX connector
Weight	50 g
Module dimensions	22 mm × 30 mm × 2.4 mm

Installing of the Network Interface Card on Jetson Nano SBC

Wireless connectivity is essential to achieve communication between IoT devices and Jetson Nano SBC. There is no in-built provision for Wi-Fi and Bluetooth connectivity on Jetson Nano SBC. Hence, we need to add a Network Interface Card (NIC) to provide the wireless connectivity. This can be done using the M.2 Key E expansion slot on the Jetson Nano SBC just below the SODIMM Jetson Nano module. The M.2 Key E expansion slots are generally designed to interface NIC.

Here, Intel Dual Band Wireless AC 8265 NIC is used which supports 802.11ac Wi-Fi dual-band with speed up to 867 Mbps and Bluetooth 4.2 connectivity. Since it is dual-band NIC, it can work at 2.4 and 5 GHz. It is the fourth-generation Intel 802.11ac NIC with 2 × 2 Wi-Fi and Bluetooth and multi-user MIMO downlink. Two 6 dBi 2.4/5 GHz Dual Band Wi-Fi RP-SMA antenna with UFL connector and IPEX cable are required for the complete setup. The specification of the Intel 8265AC is given in Table 2.2 (Intel , "Intel Dual Band Wireless AC-8265"). The steps involved in the setting up the NIC are given below.

Step 1: The Jetson Nano SBC consists of the evaluation board and the SODIMM Jetson Nano module. The SODIMM module can be identified with the heatsink present on it. Remove the SODIMM Jetson Nano module carefully by removing the two screws on either side of the module. The Jetson Nano SBC without the SODIMM module is shown in Fig. 2.25. The NIC goes below the SODIMM module.

Step 2: Now connect the antennas to the card. Connect the UFL connector from the antenna on the NIC. This task is quite difficult due to the small size of the connector and care must be taken not to accidentally damage the NIC.

Step 3: Insert the card in the M.2 Key E expansion slot just below the SODIMM slot. Fasten the screws meant to hold the NIC in place. Figure 2.26a shows the NIC screwed in place after the SODIMM is taken off.

Step 4. The next step is to insert the SODIMM module back in place. Place back the screws meant to keep the SODIMM Jetson Nano module. Figure 2.26b shows the

Fig. 2.25 Jetson nano developer kit without the SODIMM module

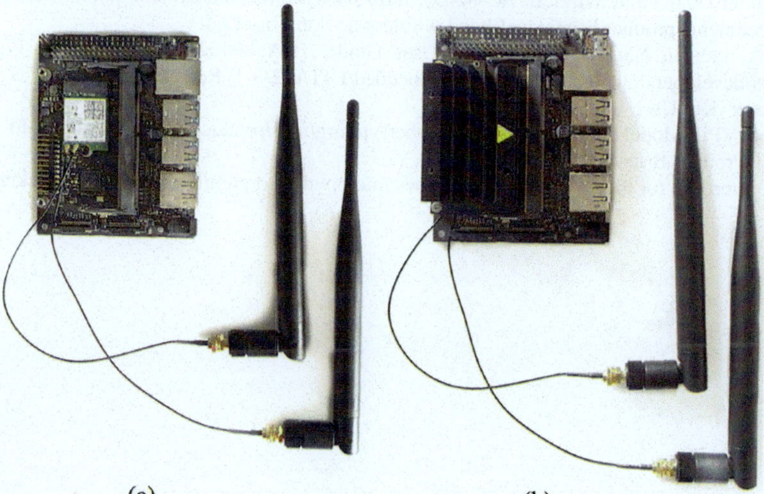

(a) (b)

Fig. 2.26 Jetson nano development board **a)** after NIC screwed in place, **b)** after the SODIMM module is screwed in place

image of Jetson Nano after the installation. The installation of the NIC is complete and the Jetson Nano can be now powered up for the first boot with wireless connectivity.

Conclusion:

In this chapter, Python is explored with three different embedded platforms such as Raspberry Pi, MicroPython Py, and Jetson Nano. Raspberry Pi is single-board computer which acts as mini personal computer. Various Raspberry Pi models are briefly compared and detailed features of Pi 3 B+ model are discussed. Two different methods of OS flashing on Pi are given in detail. Also the detailed steps involved to boot the Pi board first time are provided.

The microPython Pyboard is a small form factor board designed specifically to run MicroPython to realize various IoT projects. The detailed feature specification of Pyboard is listed. Configuration of MicroPython Pyboard is not an easy job. The detailed configuration steps of MicroPython Pyboard are provided starting with loading the firmware on Pyboard.

Jetson Nano SBC is a powerful AI platform for embedded and machine learning applications. Here the detailed board specifications are listed. Handling the Jetson Nano is not trivial and hence, the book provides the detailed steps for board configuration and setup.

References

Intel, "Intel Dual Band Wireless AC-8265". https://www.intel.com/content/dam/www/public/us/en/documents/product-briefs/dual-band-wireless-ac-8265-brief.pdf

NVIDIA, "Jetson Nano Developer Kit User Guide", DA_09402_003, December 17, 2019. https://developer.download.nvidia.com/embedded/L4T/r32-3-1_Release_v1.0/Jetson_Nano_Developer_Kit_User_Guide.pdf

"Raspberry Pi 3 Model B+". https://static.raspberrypi.org/files/product-briefs/Raspberry-Pi-Model-Bplus-Product-Brief.pdf

"Quick reference for the Pyboard". https://docs.micropython.org/en/latest/Pyboard/quickref.html

Chapter 3
Simple Applications with Raspberry Pi

Abstract All over the world, people use the Raspberry Pi to learn the coding skills, build simple projects, implement simple clusters and a little bit of Edge computing. Raspberry Pi is considered as a low-cost pocket-sized minicomputer used in various applications. In this chapter, we have focused on hands on implementation of few simple applications such as, Interfacing LEDs, Organic LED (OLED) display, Camera Interfacing to capture video/pictures, different types of motor control and lastly implementation of Bluetooth with mobile .To implement these examples, we have used Thonny Python IDE which is already integrated with Raspbian OS.

Keywords LED · PWM · Camera interfacing · Motor control · Bluetooth

3.1 Blinking of LED

To interface the LEDs with the Raspberry Pi through General-Purpose Input/Output (GPIO) pins, we need to understand how to access the Raspberry Pi GPIO.

Raspberry Pi GPIO Access:

GPIO pins of Raspberry Pi can be used to interface with the general-purpose input/output (I/O) devices. Raspberry Pi 3 B+ has on-board 26 programmable GPIO pins to interface and control many I/O devices. It can connect to the internet using on-board WiFi or WiFi USB adapter and also some pins are multiplexed as I2C, SPI, UART, etc. There are few following options available to specify GPIO pins of Raspberry Pi as shown in Fig. 3.1.

i. Physical (BOARD): Pin number from 1 to 40 corresponds to the physical location on the header.
ii. Broadcom (BCM): It is generally called "GPIO" (GPIO1–GPIO26) or RPi.GPIO
iii. WiringPi—Fig. 3.1b shows the pin numbering in Wiring Pi

J. S. Parab et al., *Python Programming Recipes for IoT Applications*, Transactions on Computer Systems and Networks, https://doi.org/10.1007/978-981-19-9466-1_3

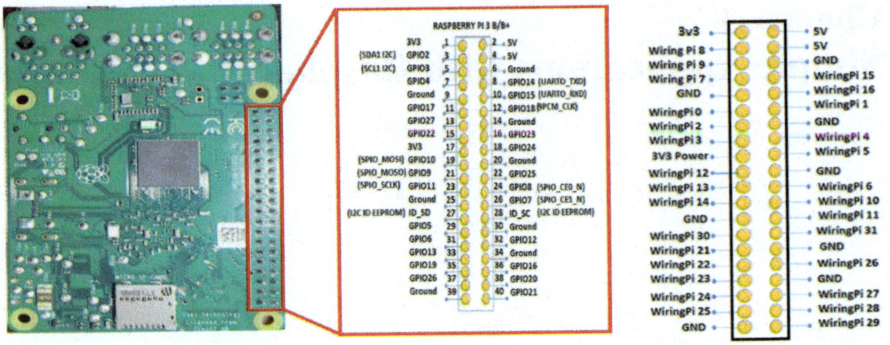

(a) (b)

Fig. 3.1 GPIO pins of physical (BOARD) and BCM. **a** Actual backside physical pin assignments and **b** WiringPi

Raspberry Pi 3 Model B+ GPIO Pins Numbering

For accessing the GPIO pin for I/O interface, Raspberry Pi has different ways of defining pin assignments from which normally two types are used, namely Physical and BCM. In GPIO Numbering (BCM), pin number refers to the number on Broadcom SoC (System on Chip), while, in Physical Numbering (BOARD), pin number refers to the pin of 40-pin P1 header on Raspberry Pi Board. The above physical numbering is simple as we can count pin number on P1 header and assign it as GPIO. To access GPIO through Python programming, first, take a simple example of how to blink the LED through Raspberry GPIO.

Installation of RPi.GPIO Python Library

The configuration of the I/O pins for read and write can be done using the 'RPi.GPIO' Python library. There are two simple methods to install the GPIO library.

Method 1: Installation of GPIO from Repository

Step 1: Open the Raspberry Pi console and update the available package versions by using the following command:

```
sudo apt-get update
```

If the package exists, then there is no need to install; otherwise, it can be installed using 'apt-get'.

Step 2: Install the RPi.GPIO package by using the following command:

```
sudo apt-get install rpi.gpio
```

If it is already installed, it will be upgraded to newer version.

Method 2: Manual Installation

The package can be downloaded from http://pypi.python.org/pypi/RPi.GPIO

Step 1: Download the library by using the following command:

```
wget    https://pypi.python.org/packages/source/R/RPi.GPIO/RPi.
GPIO-.5.11.tar.gz
```

Step 2: The downloaded file is in '.tar' format, so extract the archive to a new folder by using the following command:

```
tar -xvf RPi.GPIO-0.5.11.tar.gz
```

Step 3: Navigate to the new directory.

```
cd RPi.GPIO-0.5.11
```

Step 4: RPi.GPIO is installed using the following command:

```
sudo python setup.py installs
```

Python Program to Blink LEDs

In this example, the four LEDs are interfaced to GPIOs of Raspberry Pi. Here, one requires four LEDs, four 330 Ω resistors, and a Raspberry Pi board. Every LED has two leads—one cathode (shorter lead) and one anode (longer lead). Choose the cathode and use a 330-Ω resistor to ground it (Pin 6). The other end goes to pins 10, 11, 12, and 13, respectively. Do the connection as shown in Fig. 3.2a and the flow diagram of the entire implementation is given in Fig. 3.2b. Type the following code in Thonny IDE and execute the same.

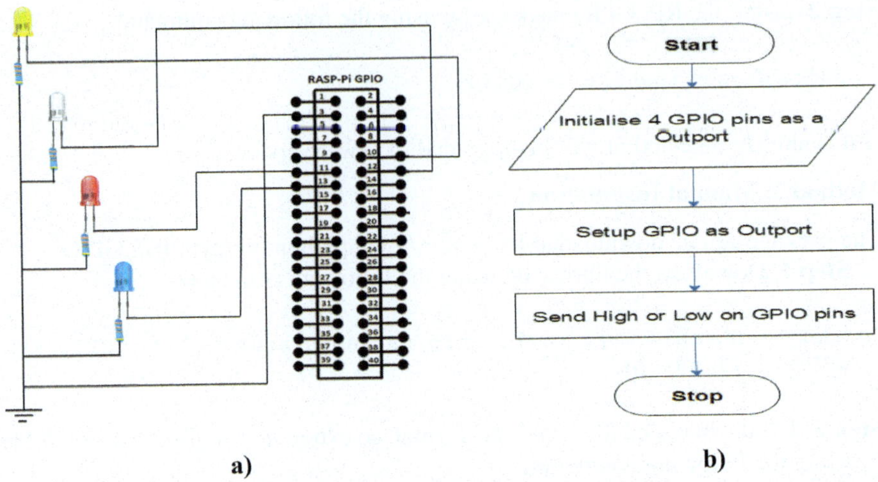

Fig. 3.2 **a** Interfacing of four LEDs with GPIO of Raspberry Pi and **b** flowchart for LED blinking

```
# To blink 4 -LED

import RPi.GPIO as GPIO
from time import sleep
GPIO.setmode(GPIO.BOARD)
pin_1=10
pin_2=11
pin_3=12
pin_4=13
GPIO.setup(pin_1,GPIO.OUT)
GPIO.setup(pin_2,GPIO.OUT)
GPIO.setup(pin_3,GPIO.OUT)
GPIO.setup(pin_4,GPIO.OUT)
GPIO.output(pin_1, HIGH)
sleep(2)
GPIO.output(pin_2, LOW)
sleep(2)
GPIO.output(pin_3, HIGH)
sleep(2)
GPIO.output(pin_4, LOW)
GPIO.cleanup()
```

Explanations of the Above Program

import RPi.GPIO as GPIO: Import RPi.GPIO package which has class to control GPIO to use Raspberry Pi GPIO pins in Python.

from time import sleep: Importing time module, when the statement "sleep (t)" is executed then the next line of code will be executed after t seconds. Example sleep (2) means the next statement will be executed after 2 s.

GPIO.setmode(GPIO.BOARD): This function is used to define pin numbering system, i.e. GPIO numbering or Physical numbering. You can also use GPIO.setmode (GPIO.BCM).

GPIO.setup(pin1,GPIO.OUT): This command is used to set the GPIO pin as the output pin.

GPIO.output(pin1, HIGH): This command is used set GPIO pin to High.

sleep (2): It will act as a delay of 2 s.

Python Program to Control LEDs using Pulse Width Modulation (PWM)

A PWM is a method for generating an analog signal using a digital source (Christ and Wernli 2014). A PWM signal consists of two main components, duty cycle and frequency. The duty cycle describes the amount of time the signal is in a high (ON) state as a percentage of the total time it takes to complete one cycle. The frequency determines how fast the PWM completes a cycle, i.e. 100 Hz would be 100 cycles per second), and therefore how fast it switches between high and low states. PWM is used for controlling the amplitude of digital signals in order to control devices. A powerful benefit of PWM is that power loss is very minimal. Compared to regulating power levels using an analog potentiometer to limit the power output by essentially choking the electrical pathway, thereby resulting in power loss as heat, PWM actually turns OFF the power output rather than limiting it. Figure 3.3a shows the PWM signal with 0–100% duty cycle and Fig. 3.3b shows the circuit diagram for understanding the concept of PWM and Fig. 3.3c shows the flowchart. Type the following code in Thonny IDE and execute the same.

```
import RPi GPIO as GPIO
from time import sleep
GPIO.Setwarning(False)
GPIO.setmode(GPIO.BOARD)
pin=18                                  #  PWM output is on pin 18
GPIO.setup(pin,GPIO.OUT)
frequency=200                           # Set duty Cycle
pwm1=GPIO.PWM(pin, frequency)
pwm1.start(0)                            # 0% Duty Cycle PWM
for i in range (0, 100):
       pwm1.ChangeDutyCycle(i)
       sleep(0.02)
for i in range (100, 0, -1):
       pwm1.ChangeDutyCycle(i)
       sleep(0.02)
pwm1.stop()
GPIO.Cleanup()
```

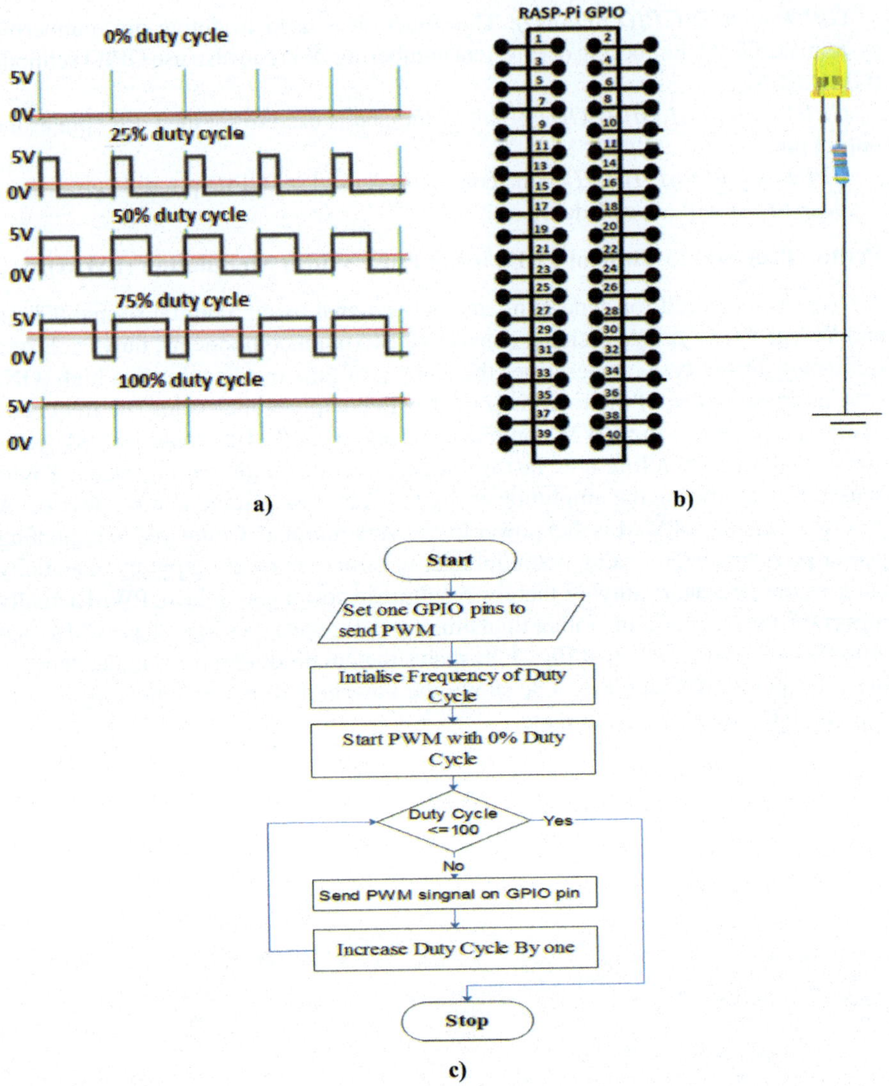

Fig. 3.3 **a** PWM signals with 0–100% duty cycle, **b** circuit for LED interfacing to Raspberry Pi, and **c** flowchart of the Python program

3.2 OLED Display Interface

To interact with the outside world and make the display more attractive as well as readable, OLED is interfaced to Raspberry Pi Board. It supports I2C communication through pins 3 (SDA) and 5 (SCL). Raspberry Pi pin numbers 1 (3.3 V) and 9 (GND) are used to power up the OLED, as shown in Fig. 3.4. The hardware connection details are shown in Fig. 3.5.

Figure 3.4 shows the interfacing circuit diagram of SSD1306 I2C OLED display with the Raspberry Pi. Power supply 3.3 V (pin 1) is connected to VCC pin of the OLED display. The SDA of the Raspberry Pi is connected to the OLED display's SDA pin. The SCL is connected with the OLED display's SCL pin. The GND of the Raspberry Pi is connected with the GND pin of the OLED display. The following steps are required to display message on OLED display.

Step 1: Connect power supply to the Raspberry Pi.

Step 2: I2C set up on Raspberry Pi:

Check if the I2C bus is activated on Raspberry Pi. Click on Raspberry Pi icon(left side corner) → preferences → Raspberry Pi configuration, as shown in Fig. 3.6a. After clicking on Raspberry Pi configuration, window will open as shown in Fig. 3.6b. The configuration window has the System, Display, Interfaces, Performance, and Localization tabs. Click on the interfaces tab and enabled the I2C (Fig. 3.6b).

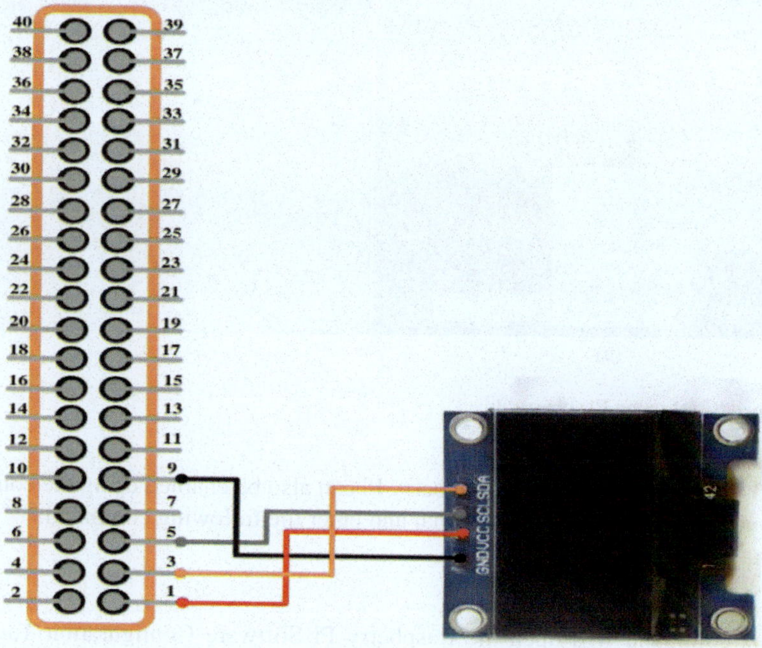

Fig. 3.4 OLED interfacing with the Raspberry Pi

Fig. 3.5 Hardware connections for OLED

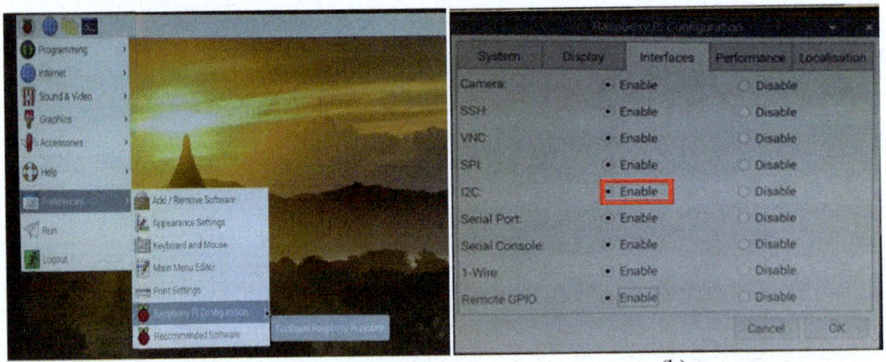

(a) (b)

Fig. 3.6 I2C set up on Raspberry Pi

Step 3: The I2C activation on Raspberry Pi can also be enabled using the command terminal. Open the command terminal and enter the following command:

```
sudo raspi-config
```

This command will open the Raspberry Pi Software Configuration Tool. Use arrow keys to scroll down and select Interface options as shown in Fig. 3.7a and press enter → the pop-up window will open as shown in Fig. 3.7b → then select the interface options → I2C option as shown in Fig. 3.7c. After this, message will be displayed as shown in Fig. 3.7d, then click on "yes" to enable I2C → Click "Ok" followed by clicking on option "Finish" as shown in Figs. 3.7e and 3.6f, respectively.

After enabling the I2C, the next step is to install the libraries required to access OLED module.

Step 1: Update and upgrade Raspberry Pi with help of the following command:

```
sudo apt update
sudo apt upgrade
```

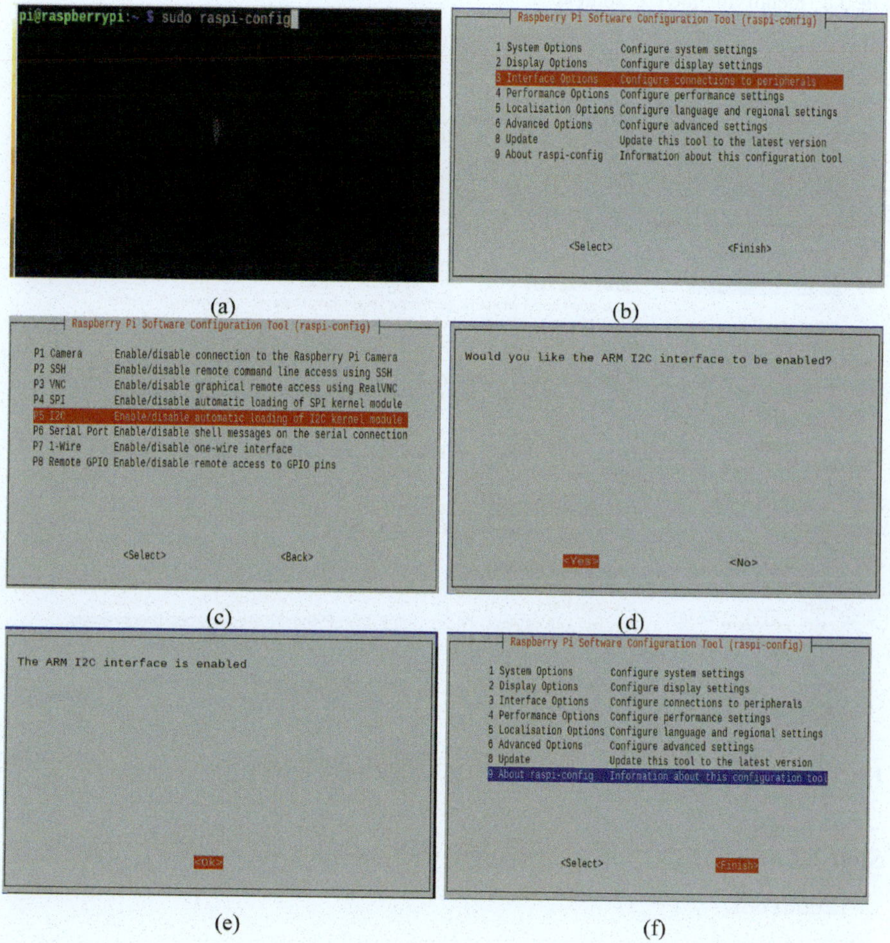

Fig. 3.7 Enabling I2C using command Prompt

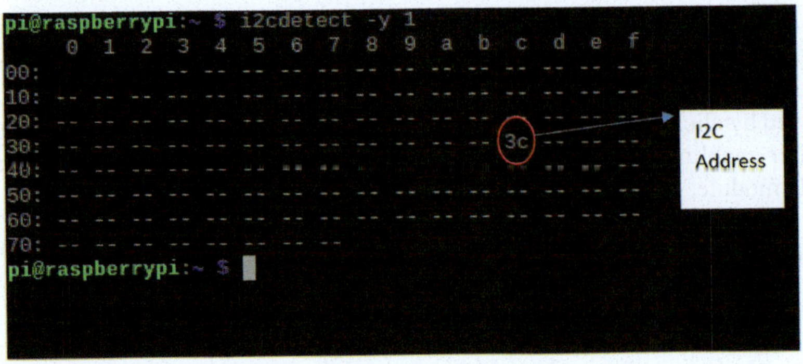

Fig. 3.8 Identifying the I2C address

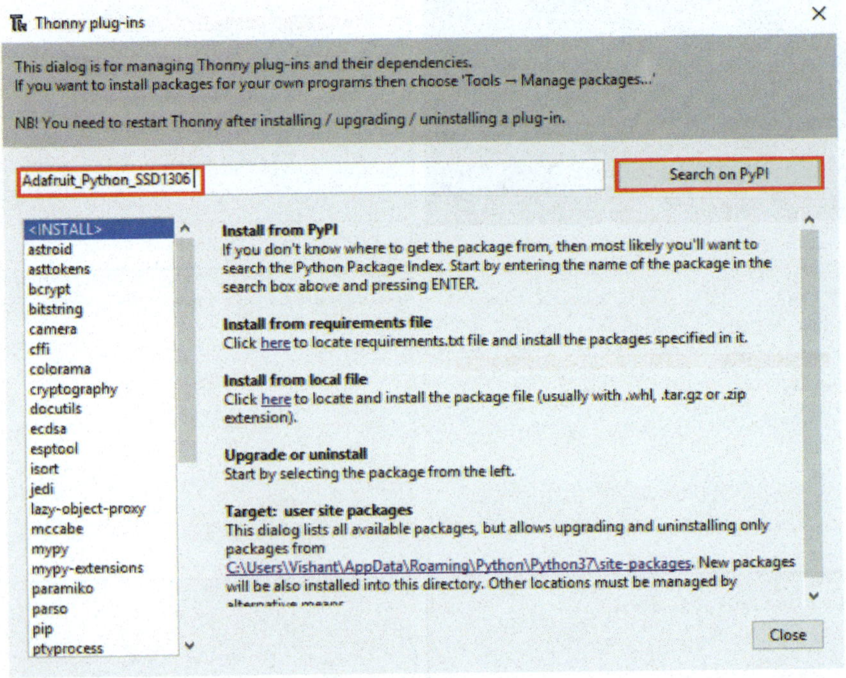

Fig. 3.9 Installation of Adafruit_Python_SSD1306 library in Thonny IDE

Step 2: Check the I2C hex address using the following command (Fig. 3.8):

```
i2cdetect -y 1
```

Step 3: Connect to the internet for installing Adafruit_Python_SSD1306 library in Thonny IDE.

Open Thonny IDE → select Tools → Manage Plug ins → enter Adafruit_Python_SSD1306 (Fig. 3.9) and then click on "*Search on PyPi*" (Fig. 3.9). The search result will be displayed and then select the library (Fig. 3.10a). Click on Install (Fig. 3.10b) to install library.

Step 4: Testing sample program on OLED

Visit https://github.com/adafruit/Adafruit_Python_SSD1306 link and click on example folder and then open shapes.py. Copy the shapes.py program and paste it in Thonny IDE followed by saving the program.

The entire flow diagram for displaying text, rectangle, etc., on OLED display is shown below in Fig. 3.11

```
# Code to display on OLED

import time
import Adafruit_GPIO.SPI as SPI
import Adafruit_SSD1306
from PIL import Image
from PIL import ImageDraw
from PIL import ImageFont

# Pi pin configuration:
RST = 24
# These lines are used with SPI:
DC = 23
SPI_PORT = 0
SPI_DEVICE = 0

# display with I2C:
disp1 = Adafruit_SSD1306.SSD1306_128_32(rst=RST)

# library initialization.
disp1.begin()

# display clear  function.
disp1.clear()
disp1.display()
```

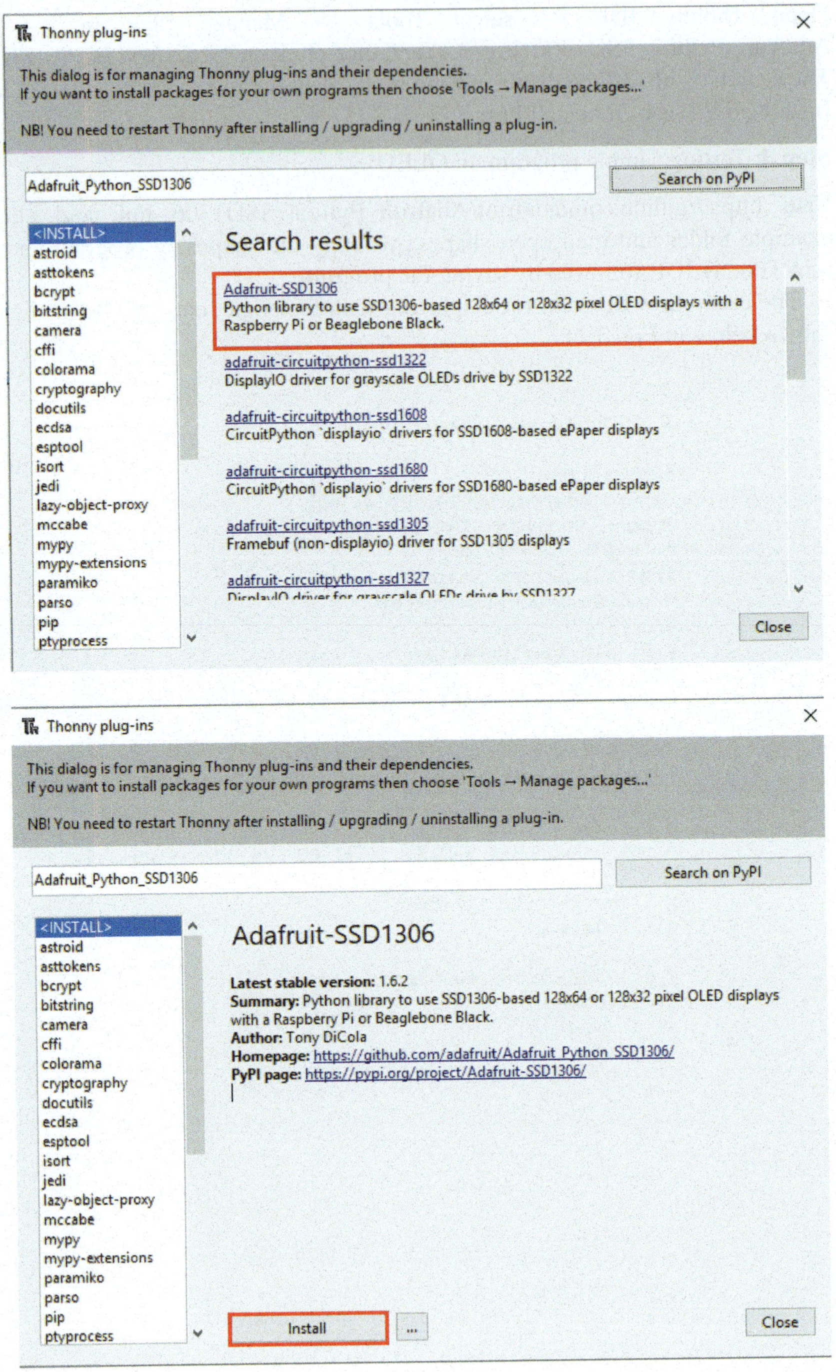

Fig. 3.10 **a** Selection of Adafruit_Python_SSD1306 library and **b** Installation of Adafruit_Python_SSD1306 library

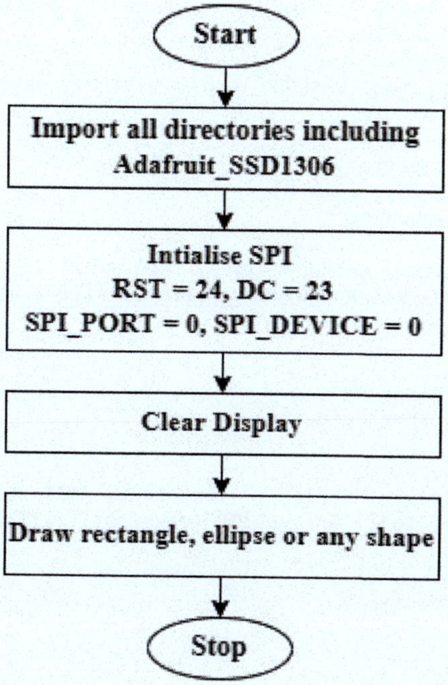

Fig. 3.11 Flowchart for displaying text, rectangle, etc., on OLED display

```
# Create blank image for drawing.
# Make sure to create image with mode '1' for 1-bit color.
width = disp1.width
height = disp1.height
image = Image.new('1', (width, height))

# draw on image with the help of drawing object.
draw = ImageDraw.Draw(image)

# clear the image with  black filled box.
draw.rectangle((0,0,width,height), outline=0, fill=0)

# Draw some shapes.
# Define  constants to allow resizing of shapes with ease.
padding = 2
shape_width = 20
top = padding
bottom = height-padding
# left to right movement keeping track of present x position for drawing shapes.
x = padding
```

```
# ellipse is drawn.
draw.ellipse((x, top , x+shape_width, bottom), outline=255, fill=0)
x += shape_width+padding
# rectangle is drawn.
draw.rectangle((x, top, x+shape_width, bottom), outline=255, fill=0)
x += shape_width+padding
# triangle is drawn.
draw.polygon([(x, bottom), (x+shape_width/2, top), (x+shape_width,
bottom)], outline=255, fill=0)
x += shape_width+padding
# Draw an X.
draw.line((x, bottom, x+shape_width, top), fill=255)
draw.line((x, top, x+shape_width, bottom), fill=255)
x += shape_width+padding

# default font is selected.
font = ImageFont.load_default()

#Write text on line.
draw.text((x, top),    'Hello',  font=font, fill=255)
draw.text((x, top+20), 'Friends!', font=font, fill=255)

# Display image.
disp1.image(image)
disp1.display()
```

3.3 Camera Interfacing

In this section, we will learn how to interface/connect the Raspberry Pi Camera Module to take pictures, record videos, and apply image effects. All current models of Raspberry Pi have a Camera Serial Interface (CSI) port (3.12a) for connecting the Camera Module as shown in following Fig. 3.12b.

Raspberry Pi Camera Module

There are two versions of the Camera Module:

- The standard version (Fig. 3.12b): Designed to take pictures in normal light.
- The NoIR version: Provided with Infrared filter so that one can use it along with an infrared light source to take pictures in the dark.

Installation Step of Picamera() Library is Given Below:

Step: To install picamera using 'apt', open console and enter the following commands:

```
sudo apt-get update
sudo apt-get install python-picamera # for python2
sudo apt-get install python3-picamera # for python3
```

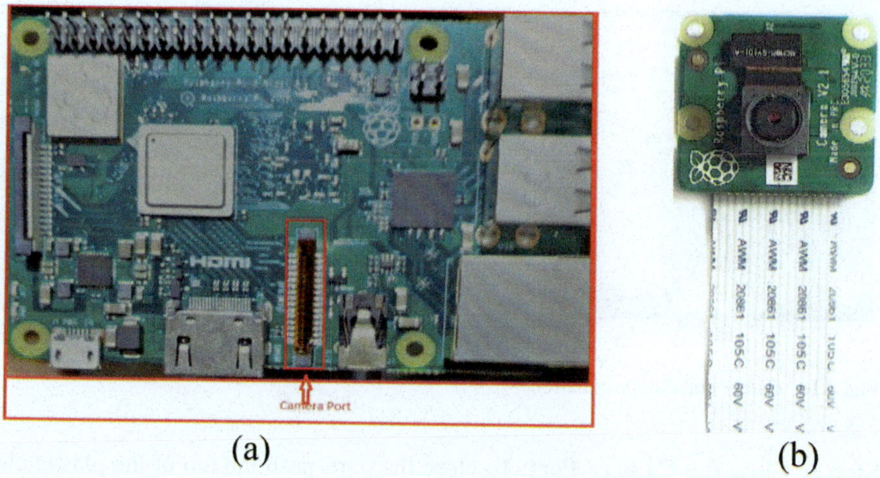

Fig. 3.12 **a** Raspberry Pi camera port and **b** camera module

One can also try the alternate method of picamera installation using Python's pip tool. The steps are given below:

Step 1: Enter the following commands to install picamera library by using Python's pip tool:

```
sudo pip install picamera
```

Step 2: Enter the following command to use the classes in the picamera array module.

```
sudo pip install "picamera[array]"
```

Step 3: Enter the following command to upgrade your installation when new releases are made:

```
sudo pip install -U picamera
```

Method to connect the Camera Module to Raspberry Pi is given below.

Step 1: Open the Camera Port on the Raspberry Pi: First ensure that Raspberry Pi is turned off and locate the Camera Module port on the Raspberry Pi 3 B+, 2 and 3, the camera port is between the LAN port and the HDMI port as shown in Fig. 3.12a. To open the port, use two fingers and lift the ends up slightly, as shown in Fig. 3.13a.

Step 2: Insert the Camera Cable: The cable has to be inserted with the right orientation with the blue side facing the Ethernet port, and the silver side is facing the HDMI port. Gently pull the clip of CSI port and insert the ribbon cable of camera module as shown in Fig. 3.13b.

(a) (b) (c)

Fig. 3.13 Camera module installation

Step 3: Close the Camera Port: To close the port, push the top of the plastic clip back into the place as shown in Fig. 3.13c.

Step 4: To enable Camera interface on Raspberry Pi: Power up the Raspberry Pi → go to the main menu → open the Raspberry Pi Configuration tool (Fig. 3.14a) → Select the Interfaces tab and ensure that the camera is enabled (Fig. 3.14b).

Step 5: Testing Camera Module Via the Command Line: Open a terminal window and enter the following command to take a still picture and save it on the Desktop (Fig. 3.14c):

```
raspistill -o Desktop/image.jpg
```

When the above command runs, you can see the camera preview opens up for 5 seconds before a still picture is taken. Look for the picture file icon on the Desktop, and double-click to open the picture.

The entire flow diagram for capturing the images using Picamera is shown below in Fig. 3.15.

Python Code to Capture Image Using Camera Module

The Python "picamera" library allows to control Camera Module.

Step 1: Open a Thonny Python IDE → Create a new file and save it as test_camera.py or any other suitable name as per your choice but make sure to avoid naming it as "picamera.py" and enter the code given below.

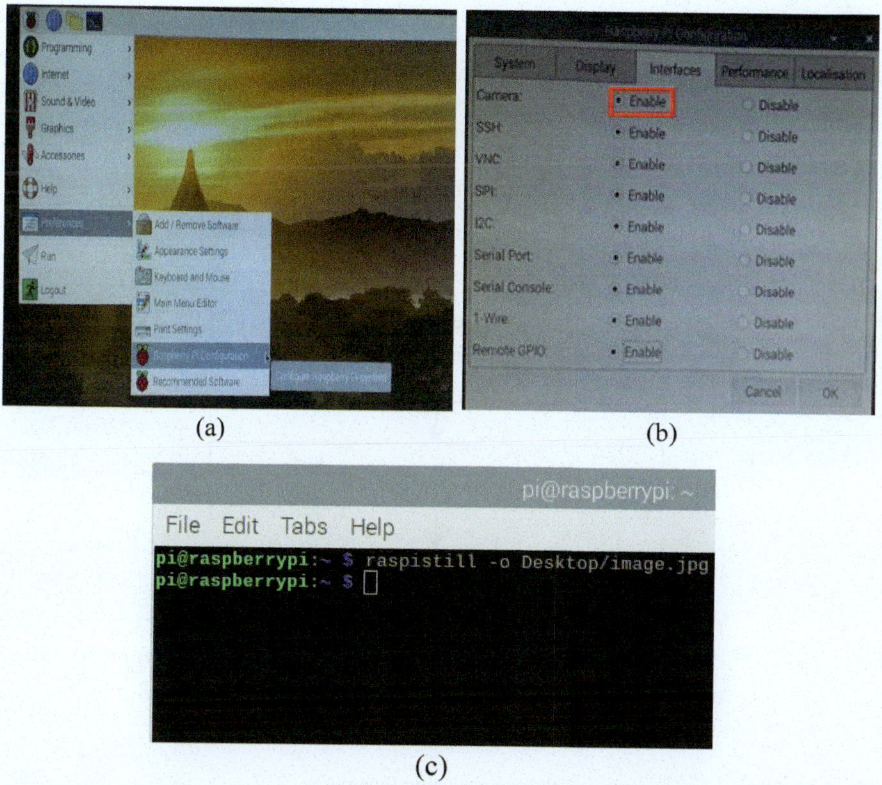

Fig. 3.14 **a,b** Camera enabling process on Raspberry Pi. **c** Terminal window to enter the command

```
#Python program for taking image by using camera module

from picamera import PiCamera
from time import sleep
filepath="/home/pi/Desktop/img%s.jpg"
camera = PiCamera()
camera.resolution=(1280,720)

for i in range (0,5):
        sleep(5)
        camera.capture(filepath %i)
print("image taken")
```

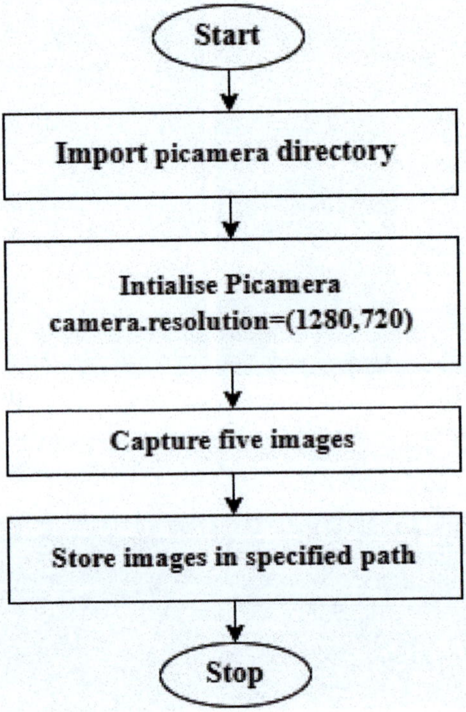

Fig. 3.15 Flowchart for capturing the images using Picamera

Step 2: Save and run the above program. The camera will take one picture every 5 s. Once the fifth image is taken, check the desktop or filepath to find the captured images (Fig. 3.16).

#Python Program to Add Text on the Captured Image

Sometimes, it is required to put some text on the captured image. An example of this can be putting a time stamp on it. Here, we have written the code which captures the image and puts the desired text on it. The flow diagram for displaying text on camera captured image is shown in below Fig. 3.17a.

Fig. 3.16 Five pictures taken by picamera module

```
#In this program we will add text "Hello Friends!" on captured image.
from picamera import PiCamera
from time import sleep
filepath="/home/pi/Desktop/image.jpg"
camera = PiCamera()
camera.resolution = (2592, 1944)
camera.framerate = 15
camera.start_preview()
camera.annotate_text = "Hello Friends!"
camera.annotate_text_size = 100
sleep(5)
camera.capture(filepath)
camera.stop_preview()
```

After executing the above code in Thonny IDE, the text is imposed on image as shown in Fig. 3.17b.

#Python Program to Record Video by Using Camera Module

We have already learned how to take images by using camera module. Now, we will see how to record the video by using the same camera module and executing the below code in Thonny IDE.

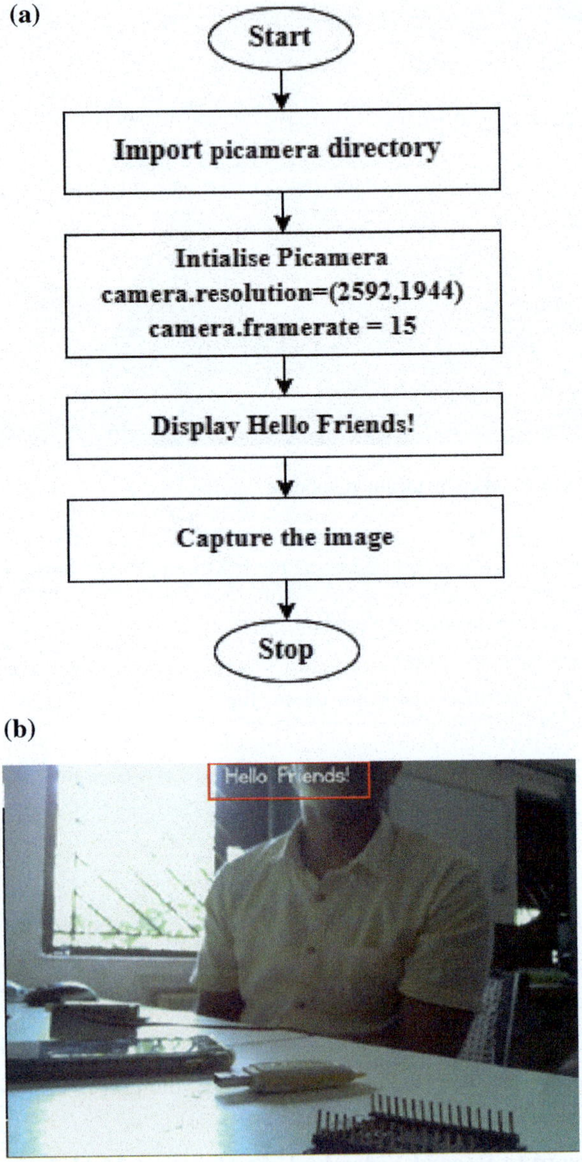

Fig. 3.17 **a** flowchart for displaying text on camera-captured image. **b** Text on picamera captured image

```
from picamera import PiCamera
from time import sleep
filepath="/home/pi/Desktop/video1.h264"
camera = PiCamera()
camera.start_preview()                    # live video preview on screen
camera.start_recording(filepath)          # Camera will start video recording
sleep(5)
camera.stop_recording()
camera.stop_preview()
```

3.4 Motor Control (DC Motor, Stepper Motor, and Servo Motor)

In this section, authors have explored various types of motor interfacing such as DC Motor, Stepper Motor, and Servo Motor with Raspberry Pi.

DC (Direct Current) Motor Control

The working principle of DC motor is that, when a magnetic field and an electric field interact, a mechanical force is produced. This is known as motoring action. There are two types of DC motors, standard and Brushless DC motors.

The DC Motor Speed Control Using the PWM Technique

The speed of DC motor is directly proportional to the supply voltage; if voltage is reduced from 12 to 6 V, then the speed becomes half that of the original speed of DC motor. But in practice, for changing the speed of a DC motor we cannot go on changing the supply voltage all the time. The voltage provided to the DC motor must be adjusted to control the speed at different torque levels (Weber 1965). The speed of the DC motor can be controlled by using PWM technique by varying the duty cycle of applied signals.

As we have already seen how to use PWM in Sect. 3.1 (control LEDs using PWM). Here, we will directly use the PWM to control the speed of DC motor. GPIO pin can source a maximum of 15 mA and the sum of currents from all 26 GPIO Pins should not exceed 50 mA. Raspberry Pi has a provision of +5 V and +3.3 V power output pins on the board for connecting other modules and sensors. This power rail is also giving power to the processor. Drawing high current from this power rail affects the processor. One can draw 100 mA safely from the +3.3 V rail. To avoid this loading effect, a separate power source is used for DC motor. The motor Driver IC L293D module is used to drive the motors. L293D is a powerful IC that can control direction and speed of two DC motors running with supply voltage ranging from 4.5 to 36 V.

Motor Driver—L293D Driver Module

L293D is a medium-power H-bridge motor driver for driving DC Motors. It consists of two H-bridge circuits for controlling each motor. H-bridge is used to change the polarity of the output, so that DC motors can be controlled in both directions. It can drive motors up to 12 V with a total DC current of up to 600 mA. The pin diagram of L293D is shown in Fig. 3.18 and pin description is as follows:

- Enable 1 and Enable 2 are the enable pins. Motors will only move if these pins are High.
- Vs is the supply voltage.
- Vss is the Logic supply voltage.
- Input1, Input2, Input3, and Input4 are the input pins.
- Output1, Output2, Output3, and Output4 are the output pins. The motors will be connected to these pins.
- GND pins are for device ground and heat sink.

Truth Table

The truth Table 3.1 provides the logical conditions to rotate motor in either clockwise or anti-clockwise direction. Table 3.2 shows the pin connections of motor with Raspberry Pi having L293 motor driver,

Figure 3.19a and b show the circuit diagram and photo of motor interfaced to Raspberry Pi, respectively. Figure 3.19c gives the flowchart for controlling DC motor. Type the below code in Thonny IDE and execute the same.

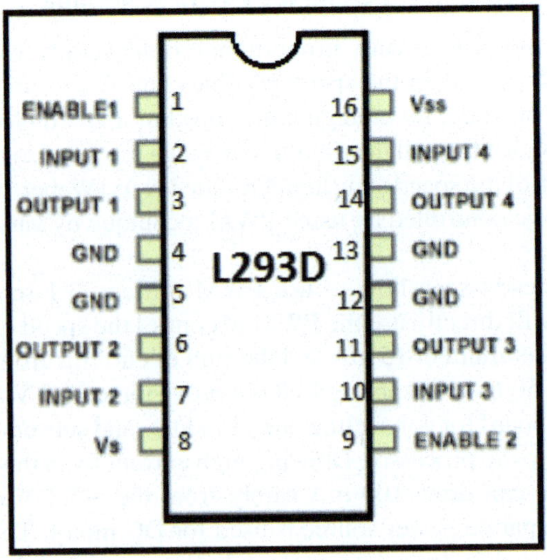

Fig. 3.18 Pin diagram of L293D

Table 3.1 Truth table for motor control

Enable	Input-1	Input-2	Output
High	High	Low	Turn anti-clockwise
High	Low	High	Turn clockwise
High	High	Low	Stop
High	Low	High	Stop
Low	X	X	X

Table 3.2 Connection of L293D motor driver module with Raspberry Pi

L293D motor driver module	Raspberry Pi
Input 1	Physical Pin 12 (GPIO18)
Input 2	Physical Pin 18 (GPIO24)
GND	GND
Enable	Physical Pin 10 (GPIO 15)

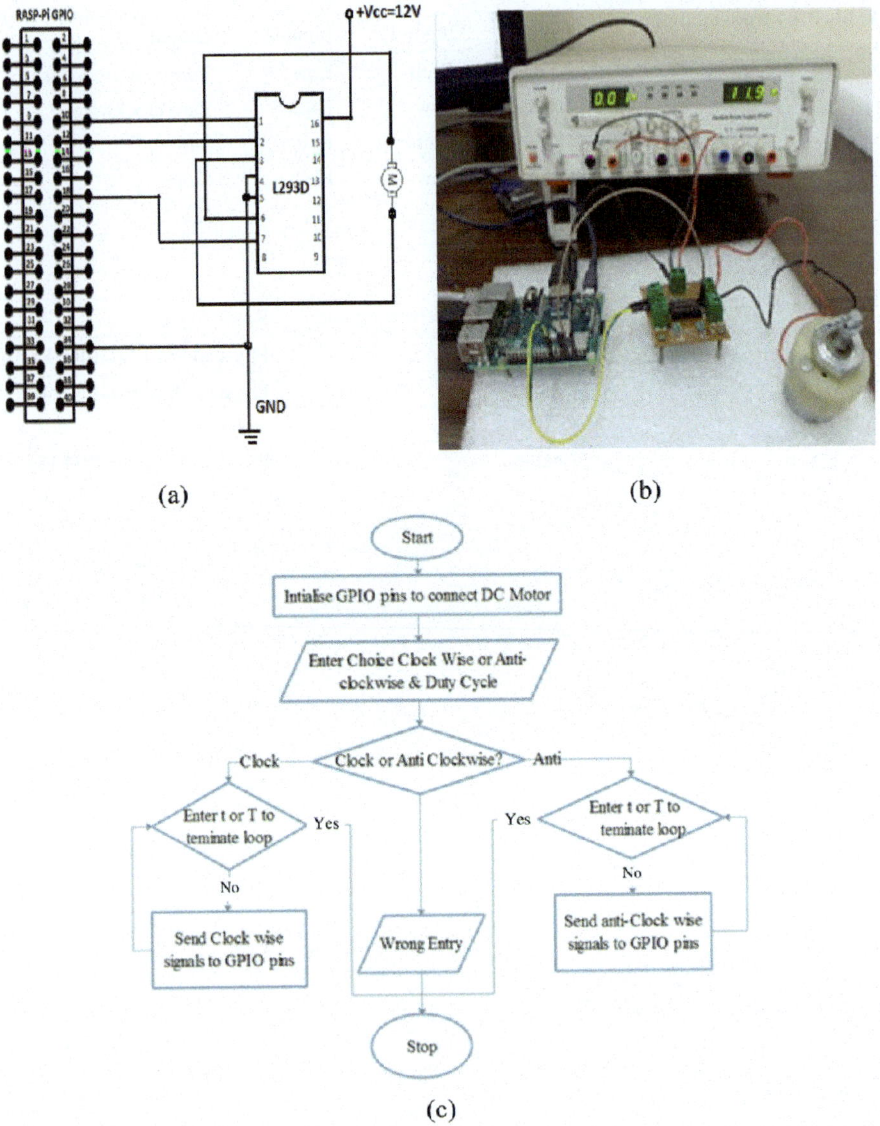

Fig. 3.19 **a** Circuit diagram for DC motor interfacing with Raspberry Pi, **b** hardware connection, and **c** flowchart

```
# Program to control DC motor

import RPi.GPIO as GPIO

from time import sleep

GPIO.setmode(GPIO.BOARD)

GPIO.setwarnings(False)

Enable1=10

input1=12

input2=18

GPIO.setup(Enable1,GPIO.OUT)

GPIO.setup(input1,GPIO.OUT)

GPIO.setup(input2,GPIO.OUT)

pwm=GPIO.PWM(Enable1,100)

pwm.start(0)

Rotation=input(" Enter c/C for clockwise & a/A for Anticlockwise:  ")
duty_cycle=int(input("Enter Duty Cycle from 1 to 100:   "))

pwm.start(duty_cycle)
delay=0.1
print("Press Ctrl+c for termination")

if Rotation=='c' or Rotation=='C':
    print ("Clockwise MOTION")
    try:
       while True :
           GPIO.output(input1,False)    # pin 2 IN1 of Driver
           GPIO.output(input2,True)   # pin 7 IN1 of Driver
           GPIO.output(Enable1,True)    # pin 1 IN1 of Driver
           sleep(delay)
      exceptKeyboardInterrupt:
           pass
```

```
elif Rotation=='a' or Rotation=='A':

    print (" AntiClockwise MOTION")

    try:

        while True:

            GPIO.output(input1,True)

            GPIO.output(input2,False)

            GPIO.output(Enable1,True)

            sleep(delay)

        exceptKeyboardInterrupt:

            pass


else:
    print("Wrong Entry")


pwm.stop()

pwm.stop()                    #stop the Pulse

GPIO.cleanup()

print("Finished")       #cleanup all of the GPIO channels.
```

Controlling Servo Motor using PWM

A servomotor is a rotary or linear actuator that enables precise control of angular or linear position, acceleration, and velocity. The main application of DC servo motors is in remote-controlled devices, robotics, and even in industrial applications (Ferrari and Ferrari 2007). Servo motors are different from ordinary motors. Depending on the specification of the servo motor, they can rotate from 0-degree to 180-degree by varying the duty cycle of the PWM signal. Servo Arm-Head moving speed can also be controlled by varying the Duty Cycle using PWM. Servo motor 90-degree position is generally referred to as "neutral" position, because it can rotate equally in either direction from that point. In this implementation, Tower pro servo motor SG90 is used. Servos can push heavy loads but they cannot lift heavy loads. PWM duration given for tower pro servo motor is 20 ms (frequency 50 Hz). PWM signal having a duration of 20 ms and signal duty cycle in between 0 to 2 ms must be generated to rotate the servo motor. Table 3.3 tabulates the duty cycle with respect to the applied current.

Table 3.3 Duty cycle with respect to the applied current

Position (degrees)	Duty cycle (ms)	Duty cycle (%)
90	1.5	7.5
180	2	12.5
0	1	2.5

PWM signal will set the angle of the servo. This can differ from servo to servo, as normally it is from 2.5 to 12.5%. To calculate the duty cycle for the desired angle, divide it by 18, then add the lowest available value, and, in this case, it is 2.5.

The formula to calculate the duty cycle is as follows:

Duty_cycle = (set_angle/18) + 2.5

So, for 90 degrees, 7.5% duty cycle and, for 180 degrees, 12.5%.

One can easily rotate the arm at a fixed angle by just varying the duty cycle. Figure 3.20a–c shows the Tower Pro SG90 9G servo Motor interfaced to Raspberry Pi with duty cycle. Figure 3.20d shows the flowchart for controlling servo motor. Type the below code in Thonny IDE and execute the same.

```
# program to control Servo motor

import RPi.GPIO as GPIO
from time import sleep
GPIO.setmode(GPIO.BOARD)
GPIO.setwarnings(False)
control_pin=10
GPIO.setup(control_pin,GPIO.OUT)  # connect to enable pin-1 of Driver IC
pwm=GPIO.PWM(control_pin,50)      # to setup the pwm commands type 50Hz freqency
set_angle=int(input(" Enter Servo Motor Rotation Angle:  "))
pwm.start(0)
delay=1
duty_cycle=(set_angle/18)+2.5
GPIO.output(control_pin,True)
pwm.ChangeDutyCycle(duty_cycle)
#GPIO.output(control_pin,True)
sleep(delay)
GPIO.output(control_pin,False)
pwm.ChangeDutyCycle(0)
pwm.stop()
GPIO.cleanup()
```

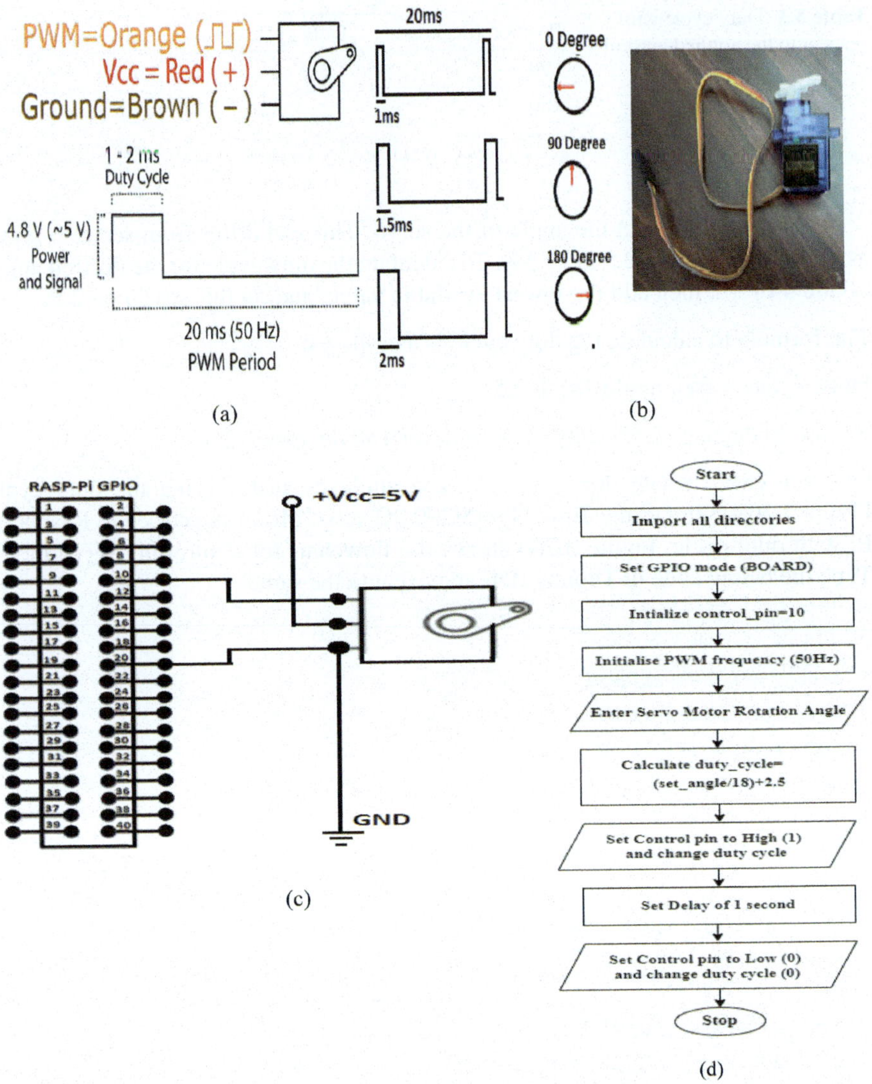

Fig. 3.20 **a** Tower Pro SG90 9G servo Motor Duty Cycle and PWM Period, **b** duty cycle with respect to the angle, **c** circuit diagram of servo motor interfacing with Raspberry Pi, and **d** flowchart for controlling servo motor

Stepper Motor Control

Stepper Motor

A stepper motor is a device that translates a DC voltage pulse train into a mechanical rotation of its shaft in a proportionate manner (Fraser 1994). Stepper motor is made up of mainly two parts, a stator and a rotor. Stator is of coil winding and rotor is mostly permanent magnet or ferromagnetic material. Stepper motors are generally used for position control. One of the best things about these motors is that they can be positioned accurately and one "step" at a time. They are a special type of brushless motor that divide a full rotation into a number of equal "steps". They are usually found in desktop printers, 3D printers, CNC milling machines, and anything else that requires precise positioning control. The speed of rotation depends upon the rate at which the control signals (Duty Cycle) are applied. A driver IC ULN2003 is used to drive the stepper motor as GPIOs of Raspberry Pi are not able to provide sufficient drive current.

In the stepper motor, continuous and limited angle rotation is obtained by providing sequential steps. Mostly, there are two step sequences, i.e. full step and half step used to rotate stepper motor as shown in Fig. 3.21. In the full-step mode, at a time two coils are excited, while, in the half-step sequence, motor moves half of its basic step angle. Tables 3.4 and 3.5 tabulate the full-step mode and half-step mode coil energizing sequence, respectively.

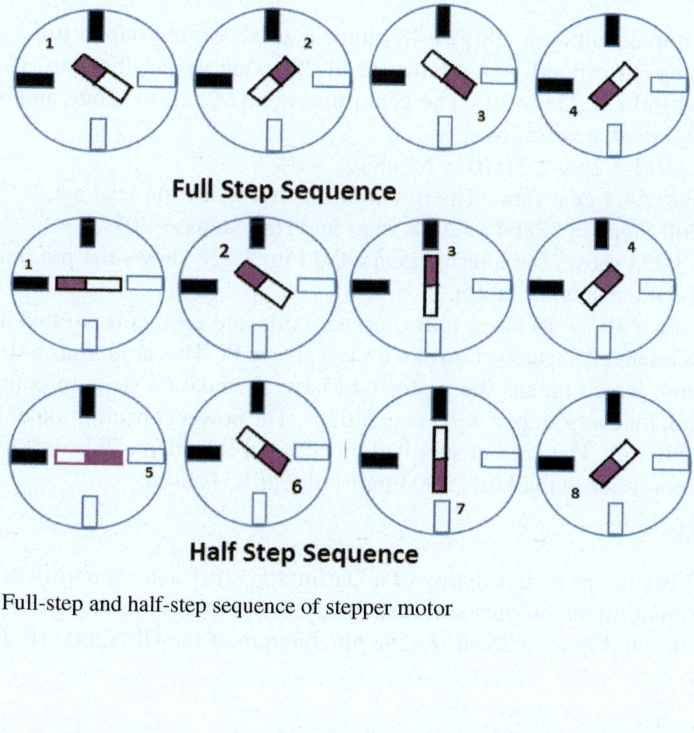

Fig. 3.21 Full-step and half-step sequence of stepper motor

Table 3.4 Full-step mode

Full-step mode				
Step	A	B	C	D
1	1	0	0	1
2	1	1	0	0
3	0	1	1	0
4	0	0	1	1

Table 3.5 Half-step mode

Half-step mode				
Step	A	B	C	D
1	1	0	0	1
2	1	0	0	0
3	1	1	0	0
4	0	1	0	0
5	0	1	1	0
6	0	0	1	0
7	0	0	1	1
8	0	0	0	1

In this implementation, 28BYJ-48 motor is used, which runs in full-step mode, and each step corresponds to a rotation of 11.25°. That means, there are 32 steps per revolution (360°/11.25° = 32). The gear ratios are 32/9, 22/11, 26/9, and 31/10.

Multiplying all gear ratios:

32/9 * 22/11 * 26/9 * 31/10 = 63.68395 ~ 64.

This gives 64:1 gear ratio. The motor has a 1/64 reduction gear set.

Total Full Steps = 32*64 = 2038 steps and Half steps = 4076.

Figure 3.22a shows the internal gears and Fig. 3.22b shows the pin number and wire colors of the stepper motor.

The motor 28BYJ-48 has a four unipolar coils and each coil is rated at +12 V; hence, it is relatively easy to control with Raspberry Pi. This motor has a stride angle of 5.625°/64, which means the motor will have to make 64 steps to complete one rotation and, for every step, it will cover 5.625°. The power consumption of the motor is around 240 mA. The current supplied by GPIO of Raspberry Pi is not sufficient to drive the motor hence the ULN 2003 motor driver IC is used.

ULN 2003

ULN2003 is a driver IC consisting of a Darlington array and capability of handling seven different inputs/outputs simultaneously. It operates in the range of 500 mA–600 mA current. Figure 3.25 shows the pin diagram of the ULN2003 (Fig. 3.23).

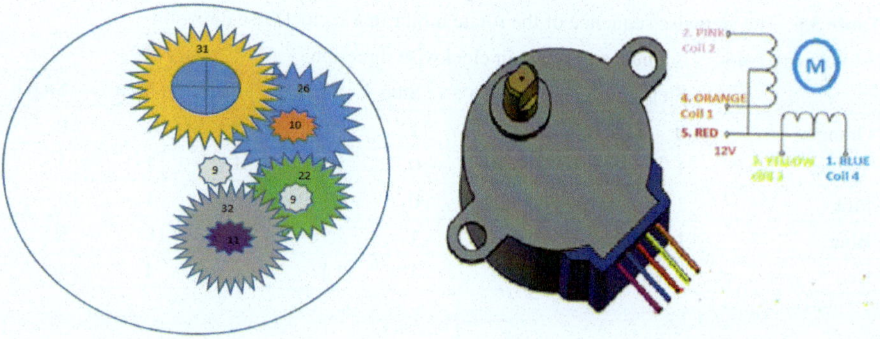

(a) (b)

Fig. 3.22 **a** internal gears of the stepper motor and **b** pin number and wire colors of the stepper motor

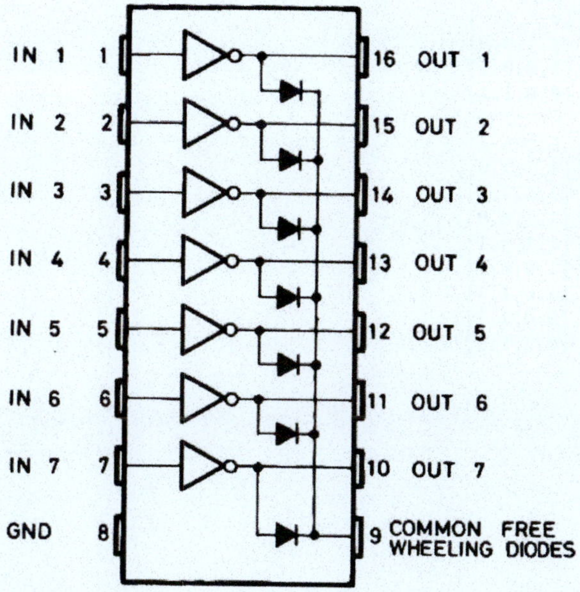

Fig. 3.23 Pin diagram of the ULN2003

Table 3.6 tabulates the stepwise sequence of the rotate motor in the clockwise direction. Figure 3.24 shows (a) circuit diagram, (b) photo of hardware connection, and (c) flowchart to control stepper motor.

Type the below code in Thonny IDE and execute the same to control stepper motor.

Table 3.6 The stepwise sequence of the rotate motor in the clockwise direction

Motor wire color	Sequence to rotate in clockwise direction							
	Step 1	Step 2	Step 3	Step 4	Step 5	Step 6	Step 7	Step 8
Orange	0	0	1	1	1	1	1	0
Yellow	1	0	0	0	1	1	1	1
Pink	1	1	1	0	0	0	1	1
Blue	1	1	1	1	1	0	0	0
Red	1	1	1	1	1	1	1	1

```
# Program for Stepper motor control
import RPi.GPIO as GPIO
from time import sleep
GPIO.setmode(GPIO.BOARD)
GPIO.setwarnings(False)

pins=[10,11,12,13]
# Full step Clock wise sequence reverse can be used for anticlockwise
full_step=[[0,1,1,0],
           [1,1,0,0],
           [1,0,0,1],
           [0,0,1,1]]

# half step Clock wise sequence reverse can be used for anticlockwise
half_step=[[0,1,1,0],
           [1,1,1,0],
           [1,1,0,0],
           [1,1,0,1],
           [1,0,0,1],
           [1,0,1,1],
           [0,0,1,1],
           [0,1,1,1]]
```

```
# To reset all pins
for pin in pins:
    GPIO.setup(pin,GPIO.OUT)
    GPIO.output(pin,0)

seq=input("Enter f/F for full step & h/H for half Step: ")

rotation=input("Enter c/C clockwise & a/A for Anticlock wise: ")
steps=512
delay=0.01
if seq=='f' or seq=='F':
    if rotation=='c' or rotation == 'C':
        for i in range(0,steps):
            for fullstep in range(0,4):
                for pin in range(0,4):
                    GPIO.output(pins[pin],full_step[fullstep][pin])
                    sleep(delay)


    elif rotation=='a' or rotation == 'A':
        for i in range(0,steps):
            for fullstep in range(3,0,-1):
                for pin in range(0,4):
                    GPIO.output(pins[pin],full_step[fullstep][pin])
                    sleep(delay)
    else:
        print("Wrong Entry")


elif seq=='h' or seq=='H':
    if rotation=='c' or rotation == 'C':
        fori in range(0,steps):
            for halfstep in range(0,8):

                for pin in range(0,4):
                    GPIO.output(pins[pin],half_step[halfstep][pin])
                    sleep(delay)
    elif rotation=='a' or rotation == 'A':
        for i in range(0,steps):
            for halfstep in range(7,0,-1):
                for pin in range(0,4):
                    GPIO.output(pins[pin],half_step[halfstep][pin])
                    sleep(delay)
    else:
        print("Wrong Entry")
else:
    print("Wrong Entry")

print("finished")
```

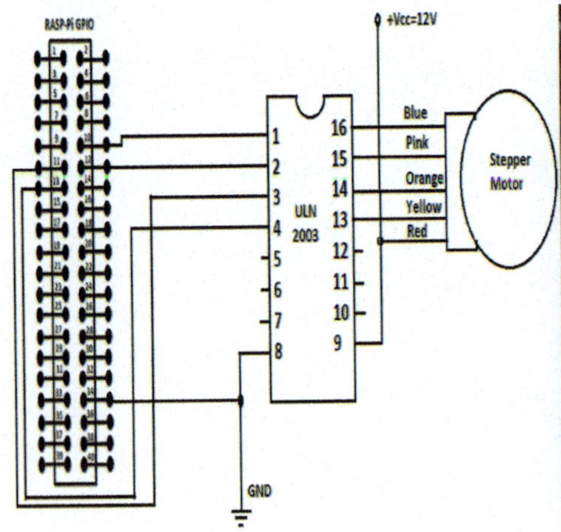

(a)

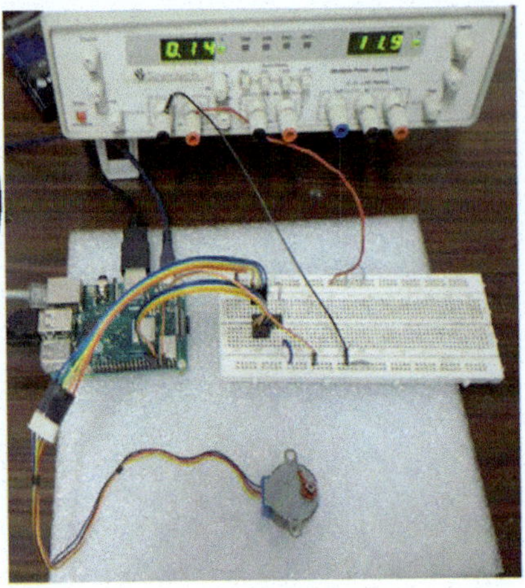

(b)

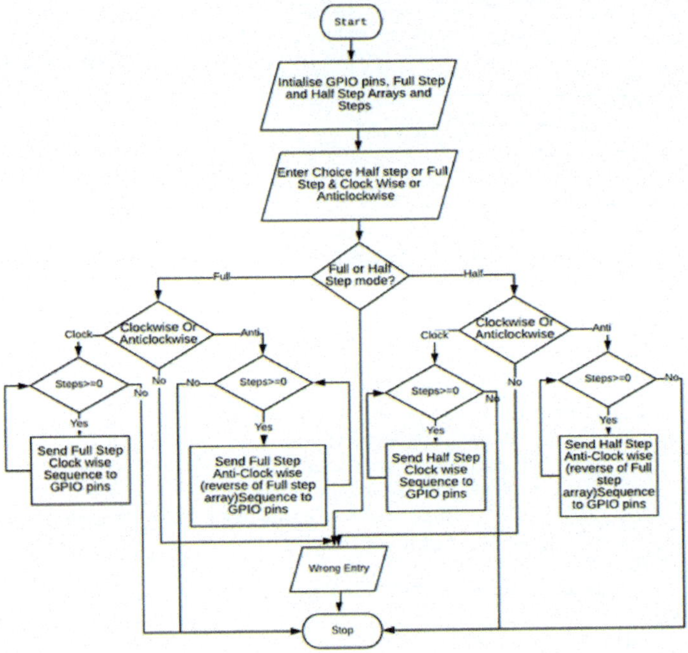

(c)

Fig. 3.24 **a** circuit diagram of the stepper motor connected with Raspberry Pi, **b** hardware connection of motor and Raspberry Pi, and **c** flowchart of the program

3.5 Raspberry Pi and Mobile Interface Through Bluetooth

Bluetooth is a wireless alternative to many of the wired communication that we use to transport voice and data (Ramandeep Kaur2, Manpreet Kaur3, J. K. 2017). Raspberry Pi 3 B+ has BCM43438 integrated chip which includes 2.4 GHz WLAN, Bluetooth, and FM receiver. The main purpose of using Bluetooth is to free up the on-board GPIO ports. Here, we have established Bluetooth communication between Raspberry Pi and smartphone to control devices.

Configuration of On-board Bluetooth of Raspberry Pi 3B+:

Raspberry Pi has an on-board Bluetooth which can be used for sending/receiving files to/from smartphone. Pairing a Bluetooth device on Raspberry Pi is same as that of a smartphone or Laptop.

Step 1: Turn-ON Bluetooth → make discoverable (Fig. 3.25a).

Step 2: Turn on Bluetooth of smartphone. Simultaneously, Select Add Device on Raspberry Pi (Fig. 3.25a). After selecting Add device, we can see mobile Bluetooth device, e.g "Samsung M20" → Select device and then click on pair as shown in Fig. 3.25b.

Step 3: Following the above step prompts for confirming the pairing code sent on the smartphone as shown in Fig. 3.25c. After the device accepts the connection by using the pair option, Raspberry Pi and the Bluetooth device will be paired and connection

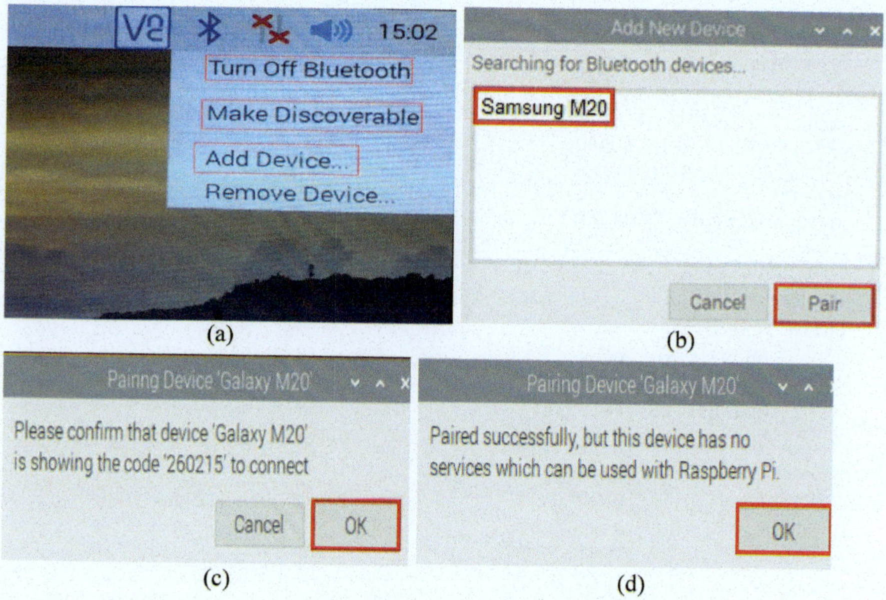

Fig. 3.25 Bluetooth connection setting between mobile and Raspberry Pi

is established as shown in Fig. 3.25d. Now the Raspberry Pi is ready to communicate via Bluetooth.

Controlling Device by Using Smartphone Through Bluetooth (Blue Dot app)

Step 1: From the Google Play store, download the Blue Dot app on smartphone for controlling devices.

Step 2: Open a terminal and enter the following command to install dbus and bluedot packages for Python 3.

```
sudo apt install python3-dbus
sudo pip3 install bluedot
```

optionally following commands shall be used for Python 2:

```
sudo apt install python-dbus
sudo pip install bluedot
```

Step 3: Upgrade to the latest version of bluedot using the following command:

```
sudo pip3 install bluedot—upgrade
```

Figure 3.26 shows the circuit diagram of the Raspberry Pi interfaced with Relay for controlling the appliances.

#Python Program to Control the Appliances Using Bluetooth

```
import os
from bluedot import BlueDot

import RPi.GPIO as GPIO
GPIO.setmode(GPIO.BOARD)

pin=11                              # Physical Pin-11=GPIO 17
GPIO.setup(pin,GPIO.OUT)

bd=BlueDot()

while True:
        bd.wait_for_press()
        GPIO.output(pin, HIGH)      # to turn on the appliance
        bd.wait_for_release()
        GPIO.output(pin, LOW)       # to turn off the appliance
```

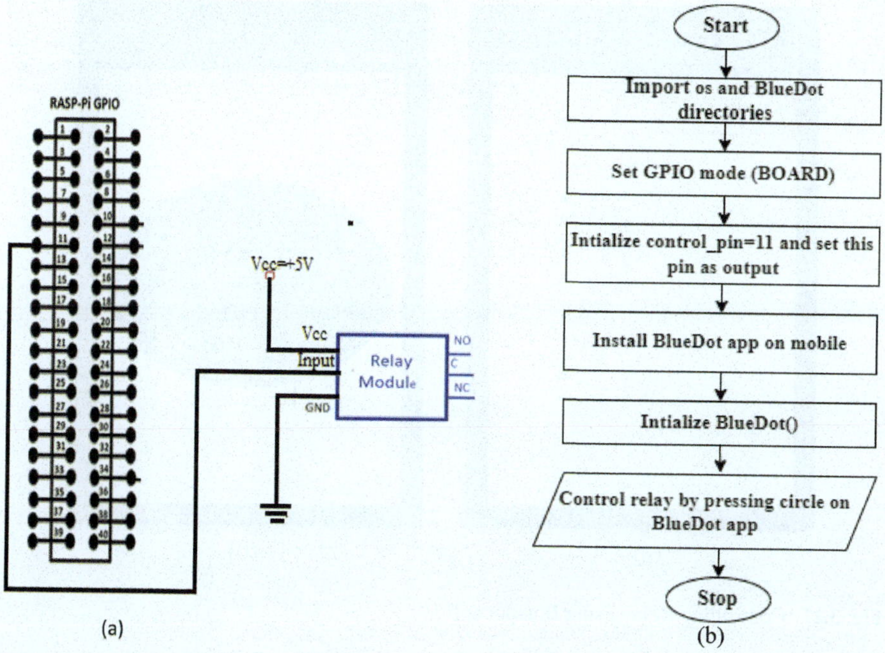

Fig. 3.26 **a** Circuit diagram of Raspberry Pi interfaced with Relay and **b** flowchart for controlling relay using BlueDot app

After executing the above program in Thonny IDE, open the Blue Dot app on your smartphone as shown in Fig. 3.27a. After connecting smartphone with Rasberry Pi, it will show a blue circle as shown in Fig. 3.27b. When the blue circle is pressed, the appliance connected to GPIO 17 (Physical Pin = 11) through relay will turn on.

Conclusion:

Raspberry Pi is an SBC installed with Raspbian OS. Various peripherals such as keyboard, mouse, and display are connected to the Raspberry Pi, which makes this system act as a mini personal computer. Raspberry Pi is popularly used for real-time Image Video Processing, IoT-based applications, and Robotics applications. In this chapter, authors have implemented Python code for controlling LEDs by using simple delay and PWM. Detailed steps involved in accessing the GPIO along with a connection diagram and execution are given. A simple I2C-based Adafruit SSD 1306 OLED is interfaced with Pi board and is also covered in a simple manner. For image and video processing, camera interfacing plays an important role. Here, authors have given the interfacing of CSI camera to Pi board in a simple manner. At the end, motor control and IoT-based home appliance control using mobile phone is covered in detail.

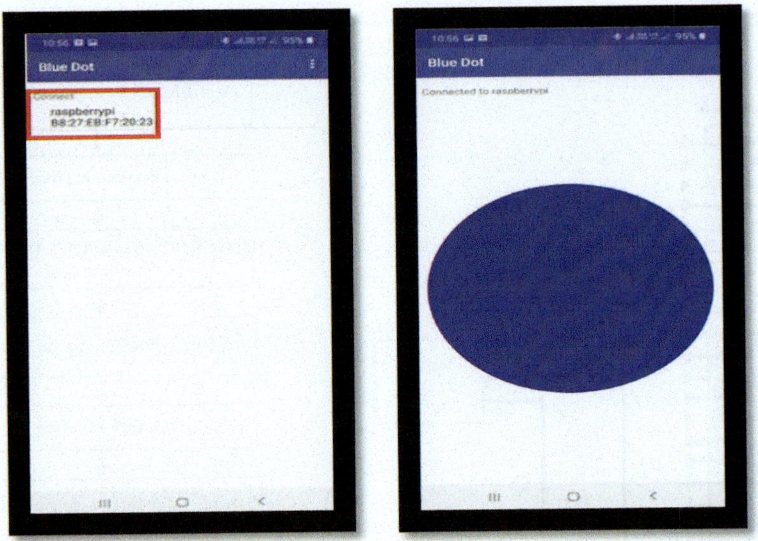

(a) (b)

Fig. 3.27 Controlling Relay using Bluedot app

Exercise:

(1) Connect 5 LEDs to Raspberry Pi. Configure the Raspberry Pi in BCM mode and write a program to blink LEDs as shown in the below pattern with an interval of 1 s between each pattern in a continuous cycle.
(2) Interface a switch and a led with Raspberry Pi. When the switch is turned ON the LED intensity should increase from Low to High and remain High. When the switch is turned OFF the intensity of LED should reduce gradually from High to Low and turn OFF(Hint: Refer to the below figure).
(3) Blink Eeight LEDs with different duty cycles using Raspberry Pi.

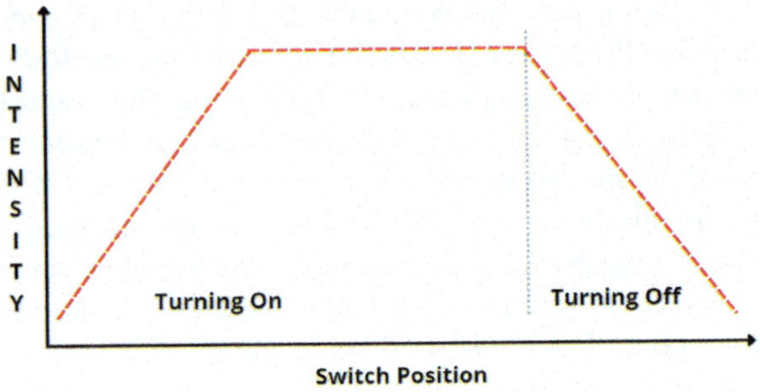

(4) Display string "Hello World", integer and floating-point number on OLED display.

(5) Interface a couple of stepper motors/DC with Raspberry Pi. Program it to run in full-step mode in a clockwise direction for 5 s followed by 10 s in anti-clockwise direction.
(6) Write Python code to control two servo motors.

References

Christ RD, Wernli RL (2014) Power and telemetry. In: Christ RD, Wernli RL (eds) The ROV manual (Second Edition), Butterworth-Heinemann, pp 141–161. https://doi.org/10.1016/B978-0-08-098288-5.00007-5

Ferrari M, Ferrari G (2007) Controlling motors. Building robots with LEGO mindstorms NXT, syngress, 2007, pp 41–59. https://doi.org/10.1016/B978-159749152-5/50008-5

Fraser CJ (1994) 2 - Electrical and electronics principles. In: Smith EH (ed) Mechanical engineer's reference book (Twelfth Edition), Butterworth-Heinemann, p 2-1-2-57. https://doi.org/10.1016/B978-0-7506-1195-4.50006-3

Kaur R, Kaur M, Kaur J (2017) Bluetooth technology. Int J Eng Comput Sci 5(3).http://www.ijecs.in/index.php/ijecs/article/view/702

Weber HF (1965) Pulse-width modulation DC motor control. IEEE Trans Ind Electron Control Instrum IECI 12(1):24–28. https://doi.org/10.1109/TIECI.1965.229545

Chapter 4
MicroPython PyBoard for IoT

Abstract In this chapter, readers will learn how to program Pyboard for IoT applications using MicroPython. MicroPython is a member of the Python interpreter's family. MicroPython Programming language is a subset of the Python to fit and execute on a microcontroller. It has several optimizations to ensure that it works effectively and consumes little RAM. Hence MicroPython is perfect for embedded systems and IoT applications. Here, authors have focused on hands on implementation of few simple IoT applications such as home automation, smart e-waste bin, industrial environmental monitoring, green house monitoring, and aquaculture monitoring. Before implementing the IoT applications, one has to learn how to Install, run and debug the Micropython using PyCharm IDE.

Keywords Home automation · Environmental monitoring · Green house · Aquaculture · e-waste bin

MicroPython Installation in PyCharm IDE

PyCharm IDE supports Pyboard, ESP8266, and BBC Micro:bit microcontroller devices. In order to use Pyboard, one has to install MicroPython plugin in PyCharm. The detailed installation and setup process of the MicroPython plugin in PyCharm is as follows:

Step1: Download and install PyCharm IDE from the link: https://www.jetbrains.com/pycharm/

Step 2: To install the MicroPython plugin in PyCharm: open PyCharm IDE → *File* → *Settings* → *Plugins* → Enter 'MicroPython' in search window → click on 'Install' button to install MicroPython plugin.

Step 3: Create Python project: Go to *File* → *New Project (e.g TestIoT)* and then click on 'Create' button.

Step 4: Setting project structure: Go to *File* → *Settings* → *Project: TestIoT* → *Project structure*. Then right-click on '.Idea' and 'venv' and select excluded.

J. S. Parab et al., *Python Programming Recipes for IoT Applications*, Transactions on Computer Systems and Networks, https://doi.org/10.1007/978-981-19-9466-1_4

Step 5: To enable MicroPython support and path selection: Go to *File* → *Settings* → *Languages & Frameworks* → *MicroPython* and select device type as Pyboard. Tick the Check boxes '*Enable Micropython support*' and '*Auto detect device path*'. Device path can be also given manually and then click on 'detect' button and then click on 'OK'.

Step 6: Restart the PyCharm IDE and right-click on project name (TestIoT) → New → Python file → Enter the name to Python file (i.e main.py).

Step 7: After naming the file one has to compulsorily install missing packages as prompted by the IDE.

Step 8: Setting Python Read Evaluate Print Loop (REPL): click on Tools → MicroPython → MicroPython REPL. This step helps to show errors and allows to run MicroPython shell on Pyboard device.

Step 9: Flash Files to Pyboard: On taskbar, go to Run → Click on 'Flash main.py'.

Testing Simple code to Turn on LED of Pyboard:

After the successful installation of PyCharm and MicroPython plugin, the IDE is ready.

To test the simple code to turn on Pyboard LED, enter and flash the code given below:

```
import pyb
pyb.LED(1).on()
pyb.LED(2).on()
pyb.LED(3).on()
pyb.LED(4).on()
```

4.1 Home Automation

Nowadays, home automation has become more popular, and smart systems are being implemented practically in every home. Home automation, often referred to as "Smart Home Technology", which uses technology to automate your home. Home automation allows you to control almost every aspect of your home through the IoT. The concept of a smart home has gained popularity in recent years as technology made life easier (Majeed et al. 2020). Almost everything has gone digital and automated (Stolojescu-Crisan et al. 2021). The IoT concept, conceptualizes the idea of remotely connecting and monitoring real-world items (things) over the internet (Madhesh et al. 2020). One of the most touted benefits of home automation is providing peace of mind to homeowners, allowing them to monitor/control their homes remotely, countering dangers such as forgetting to turn off geyser or a front door left unlocked or lights/fan left on. Figure 4.1 shows the basic smart home system.

Fig. 4.1 Smart home systems

Here, the authors implemented home automation system with two approaches, first one with Pyboard and second one with ESP32 as IoT device, using Blynk app. The reason for using ESP32 is that Pyboard is not included in Blynk app to remotely control the appliances. Here, four electrical appliances such as fan, refrigerator, lights, and air condition (AC) are controlled using IoT device. This system has ESP32 configured as IoT device to remotely control the above appliances. The electrical appliances in the house are connected through relay module to Pyboard/ESP32. The OLED display interface is used to show the status of appliances connected to the Pyboard/ESP32.

Here, a 0.96-inch OLED SSD1306 display is used. The OLED display will communicate with the Pyboard/ESP 32 via the I2C protocol.

First Approach:

Home Appliances Control with Pyboard:

The four electrical appliances (fan, AC, refrigerator, and light) will be controlled by Pyboard through the relay as shown in Fig. 4.2a. Figure 4.2b shows the entire flowchart for controlling the four appliances. Table 4.1 tabulates the connection of Pyboard with switches, OLED, and relays.

Implementation and Configuration Steps:

Step 1: Copy MicroPython OLED ssd1306 library from the following link:

https://github.com/micropython/micropython/blob/master/drivers/display/ssd1306.py

Step 2: Open PyCharm IDE → paste the code in PyCharm IDE and save it as ssd1306.py.

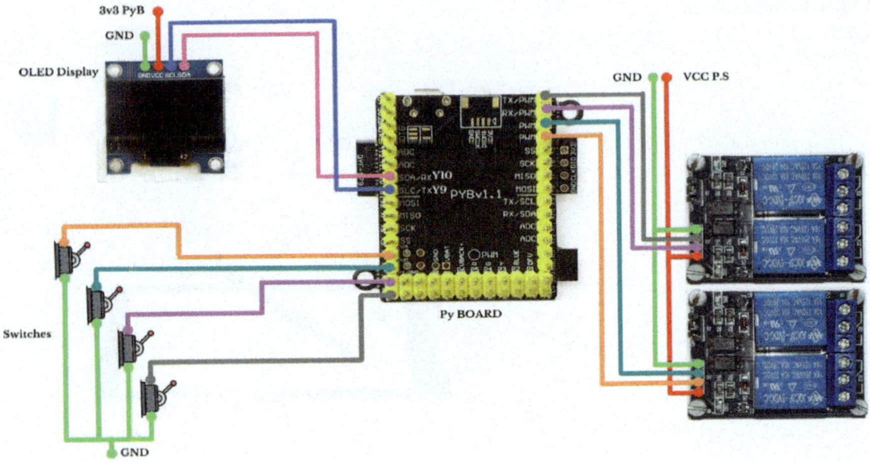

(a)

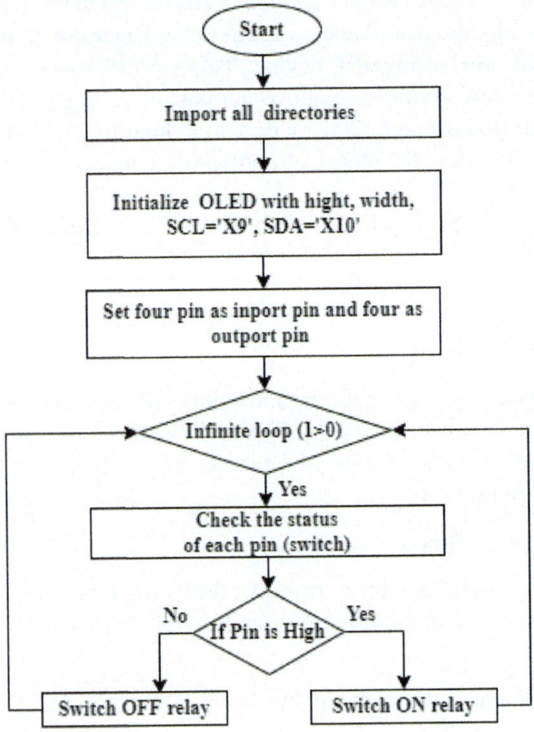

(b)

Fig. 4.2 **a** Circuit diagram of home automation and **b** Flowchart for controlling four devices

Table 4.1 Pyboard, switches, and relay pins connections

Pyboard	Switches	Pyboard	Relay	Pyboard	OLED
X1	SW1	Y1	R1	3.3 V	Vin
X2	SW2	Y2	R2	GND	GND
X3	SW3	Y3	R3	X9	SCL
X4	SW4	Y4	R4	X10	SDA

Step 3: Make sure the connections are according to Fig. 4.2 and enter the following code in 'main.py':

```
# Program to control home appliances
import time
import ssd1306
import machine
from pyb import Pin
from micropython import const
width = const (128)
height= const (64)
ssd1306_scl= Pin('X9', Pin.OUT_PP)
ssd1306_sda= Pin('X10', Pin.OUT_PP)
i2c_ssd1306=machine.I2C(scl=ssd1306_scl, sda=ssd1306_sda)
oled = ssd1306.SSD1306_I2C(width, height, i2c_ssd1306)
oled.fill(0)
p_in1 = Pin('X1',Pin.IN, Pin.PULL_UP)
p_in2 = Pin('X2', Pin.IN, Pin.PULL_UP)
p_in3 = Pin('X3', Pin.IN, Pin.PULL_UP)
p_in4 = Pin('X4', Pin.IN, Pin.PULL_UP)
p_out1 = Pin('Y1', Pin.OUT, Pin.OUT_PP )
p_out2 = Pin('Y2', Pin.OUT_PP)
p_out3 = Pin('Y3', Pin.OUT_PP)
p_out4 = Pin('Y4', Pin.OUT_PP)
while (1):
    oled.fill(0)
    if p_in1.value() == False:
        p_out1.high()
        oled.text('AC-ON', 0, 0)
        oled.show()
    else:
        p_out1.low()
        oled.text('AC-OFF', 0, 0)
        oled.show()
    if p_in2.value() == False:
        p_out2.high()
        oled.text('Fan-ON', 0, 10)
        oled.show()
    else:
        p_out2.low()
        oled.text('Fan-OFF', 0, 10)
        oled.show()
    if p_in3.value() == False:
        p_out3.high()
        oled.text('Refrigerator-ON', 0, 20)
        oled.show()
```

```
else:
    p_out3.low()
    oled.text('Refrigerator-OFF', 0, 20)
    oled.show()
if p_in4.value() == False:
    p_out4.high()
    oled.text('Light-ON', 0, 30)
    oled.show()
else:
    p_out4.low()
    oled.text('Light-OFF', 0, 30)
    oled.show()
time.sleep(0.5)
```

This is just a simple application to control the appliances using switches. To control appliance over internet, the authors have demonstrated the same by using ESP32.

Second Approach:

Home Appliances Control with ESP32 (IoT)

The four electrical appliances (fan, AC, refrigerator, and light) will be controlled by ESP32 through the relay as shown in Fig. 4.3. Table 4.2 tabulates the connection of ESP32 with relay.

To program ESP32, the Thonny IDE is used as it supports ESP32 board. The detailed steps for home automation implementation are as follows:

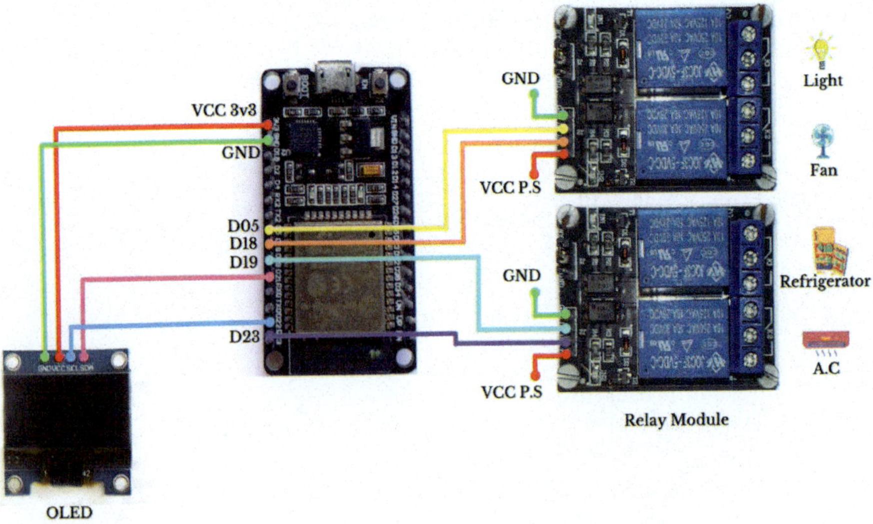

Fig. 4.3 Complete circuit diagram of home automation system

Table 4.2 ESP32 and relay pins connections

ESP32	Relay	ESP32	OLED
23	R1	3V3	VCC
19	R2	GND	GND
18	R3	D19	SDA
5	R4	D23	SCA

Step 1: Firmware installation using Thonny IDE

(1) Download the latest firmware from the link https://micropython.org/download/esp32/
(2) Installing MicroPython Firmware onto ESP32: Open Thonny IDE → Tools → options → interpreter → select MicroPython (ESP32) → select COM port → click on 'install and update firmware' → select port and firmware path (downloaded MicroPython Firmware path) → then click on install.

Now, the ESP32 is ready to use for any application after firmware installation.

Step 2: Blynk app installation process

The ESP32 board is used to control appliances with Blynk app via WiFi. Blynk app supports both iOS and Android platforms to control ESP32, Arduino, and Raspberry Pi over the internet. It is a digital dashboard, where one can build a graphic interface for project by simply dragging and dropping widgets. Follow the below steps to set up the Blynk app on mobile.

(1) Download Blynk app (legacy) from Google Play Store or App Store and install it.
(2) Create an account and log in.
(3) Create a new project and name it with a suitable name followed by selecting ESP32 device.
(4) Click on 'Create' button. You will receive an authentication key on your registered email id.
(5) Then, tap anywhere on the canvas to open the widget box. All the available widgets are located here. From the available options choose a 'button'.
(6) In this application, create four buttons to control four appliances through Blynk app. Tap on the widget to change the setting. Select the PIN → Digital → gp21. Choose button mode as 'switch'. Continue this step to create other buttons such as gp19, gp18, and gp5.
(7) Now, the Blynk app is ready. On pressing 'Play' button, it will switch from 'EDIT' mode to 'PLAY' mode where one can interact with the hardware.

Step 3: Controlling home appliances using Blynk

(1) Copy Blynk MicroPython library from https://github.com/lemariva/uPyBlynk/blob/master/BlynkLibESP32.py

(2) Edit the boot file (boot.py) of ESP32: File → open → Select 'MicroPython' → open 'boot.py' and paste the below code and save it.

```
# The boot.py file

ssid_ = "samsung"                     #Change your WiFi ssid
wp2_pass = "qwerty123"                #Change your WiFi password
def do_connect():
    import network
    sta_if = network.WLAN(network.STA_IF)
    if not sta_if.isconnected():
        print('connecting to network...')
        sta_if.active(True)
        sta_if.connect(ssid_, wp2_pass)
        while not sta_if.isconnected():
            pass
    print('network config:', sta_if.ifconfig())
do_connect()
```

(3) Enter the below code in 'main.py' file with correct authentication token received over email.

```
from machine import Pin, SoftI2C
from time import sleep
import network
import utime as time
from machine import Pin
import BlynkLibESP32 as BlynkLib   # for ESP32
# blynk = BlynkLib.Blynk("Auth token")
blynk = BlynkLib.Blynk("6GKWghlS8rPx07pQBVox63EcNif6klBA")
blynk.run()
```

The flowchart for home automation system using Blynk app is shown in Fig. 4.4a. Run the 'main.py' program and the obtained output is shown in Fig. 4.4b.

4.2 Smart e-waste Bin

In the present days, a rapid increase in urbanization and per capita income has led to an increase in municipal solid waste generation. Society creates an unhygienic environment for its citizens with respect to waste generation. This rapid generation of waste leads to various infectious diseases in the environment. A smart waste management (SWM) system ensures real-time monitoring of collection and transportation of waste. The SWM ensures that waste is collected on time and that the cost of entire operation is kept to a minimum (Zeb et al. 2019). To deal with various sorts of waste, including biological, industrial, and home waste, a variety of techniques are used (Rahman et al. 2020). Technologies such as global positioning system (GPS), radio frequency identification (RFID), global system for mobile communications (GSM), machine-to-machine (M2M) communication, and IoT, as well as innovative mobile

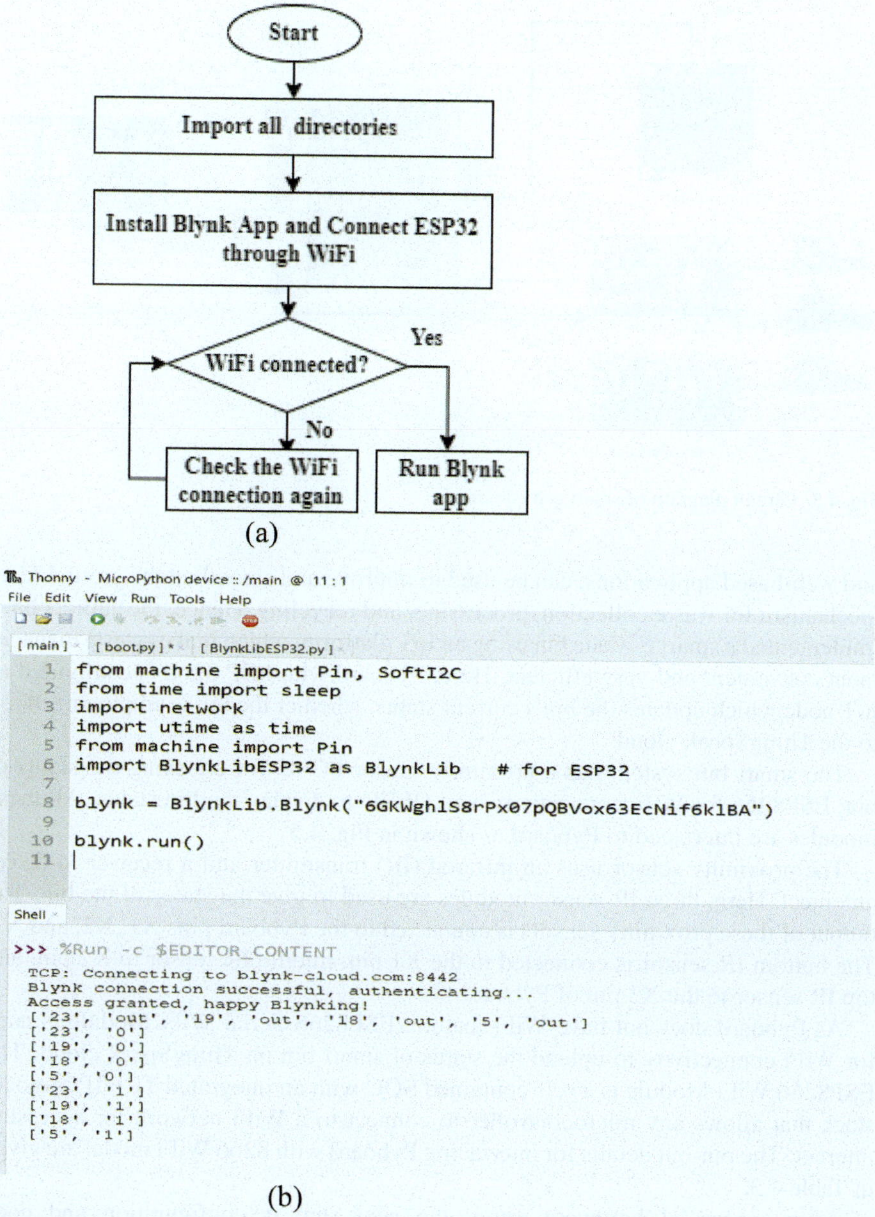

Fig. 4.4 **a** Flowchart for home automation system using Blynk app and **b** Output status of home appliance

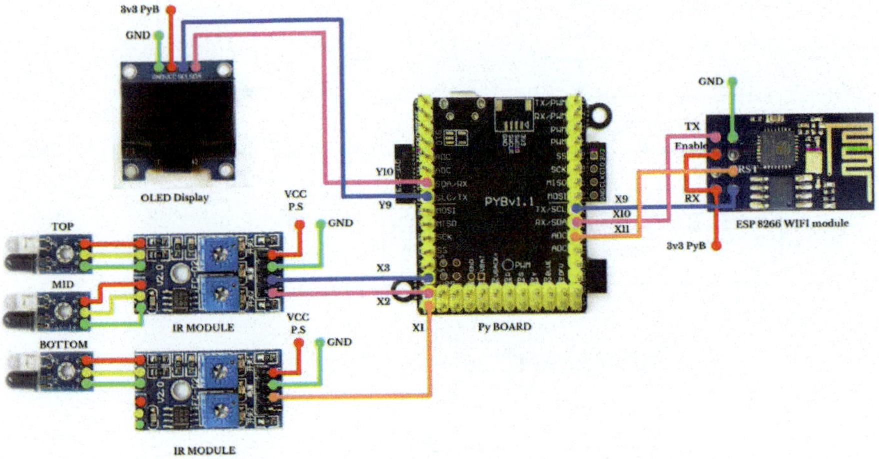

Fig. 4.5 Circuit diagram of smart e-waste bin

and web-based applications, can be used to improve and smoothen the ground-level mechanism for waste collection, processing, and recycling. Hence, the authors have implemented a smart e-waste bin using an IoT platform, which makes waste management convenient and very efficient. Here, Pyboard with ESP 8266 is configured as IoT node which updates the bin's current status, whether the bin is empty or full, on to the ThingSpeak cloud.

The smart bin system has a proximity sensor FC-45 for checking the status of bin, ESP8266 for WiFi connectivity, and OLED to display the bin status. All these modules are interfaced to Pyboard as shown in Fig. 4.5.

The proximity sensor uses an infrared (IR) transmitter and a receiver to detect an object. Here, three IR sensor modules are used to give the status of the bin. The output of these proximity sensors is connected to the Pyboard pins X1, X2, and X3. The bottom IR sensor is connected to the X1 pin, middle IR sensor to X2 pin, and top IR sensor to the X3 pin of Pyboard.

As Pyboard does not have WiFi feature, ESP8266 Serial WiFi Module is used for WiFi connectivity to upload the status of smart bin on ThingSpeak cloud. The ESP8266 WiFi Module is a self-contained SOC with an integrated TCP/IP protocol stack that allows any microcontroller to connect to a WiFi network for accessing internet. The pin-out details for interfacing Pyboard with 8266 WiFi model are given in Table 4.3.

After successful hardware setup, the next step is configuration and code implementation.

Implementation and configuration:

Table 4.3 ESP8266 WiFi module pin description and pin connection with Pyboard

8266 Pin	Description	Pyboard pins
VCC	Power pin = 3.3v	3.3 V
GND	Ground	GND
Rx	Receive serial data from another device	Tx
Tx	Transfer serial data to other devices	Rx
CH_En	Chip enable pin, connected to 3.3 V	3.3 V
GPIO 0	General-purpose input–output pin used as a normal GPIO pin and also used to enable the ESP8266 programming mode	Not connected
GPIO 2	Used as a GPIO pin	Not connected

Step 1: Configuration of ThingSpeak Channel

To create channel on ThingSpeak, one has to first sign up on ThingSpeak (https://thingspeak.com/). In case one has an account on ThingSpeak, just sign in using your id and password. For signup, fill in your details and then verify with the received e-mail and proceed. After this, click on 'New Channel' button which will subsequently ask for the 'Name and Description' of the data you want to upload on this channel as shown in Fig. 4.6

(1) In this application, smart bin status is sent to ThingSpeak. Hence, authors have named the channel 'Smart Bin'. More than one field of data can be activated by checking the box next to Field option. The authors have created one field, namely 'Bin status' (Fig. 4.7). After this, click on 'save channel' button to save the details.

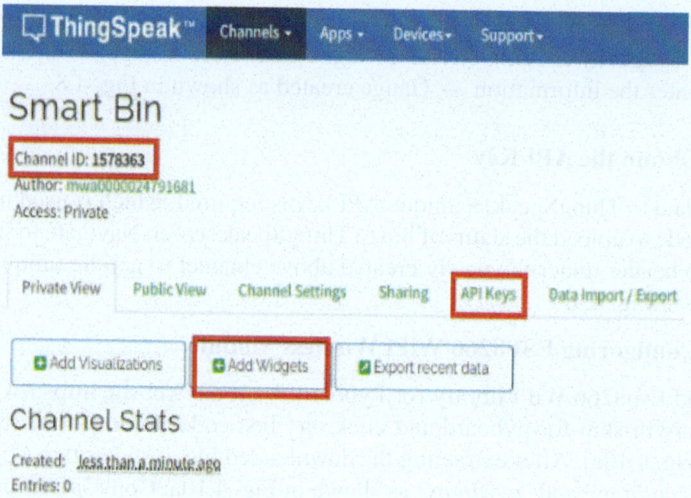

Fig. 4.6 Channel details for smart bin

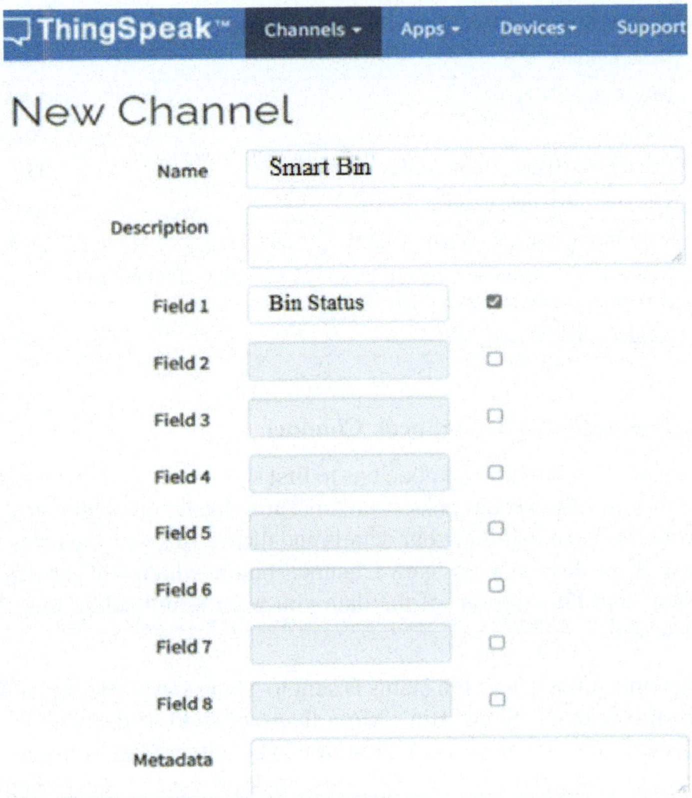

Fig. 4.7 Channel name and field details

(2) Add widget by clicking on 'Add widget' and select 'Gauge' → click on 'Next' → enter the information → Gauge created as shown in Fig. 4.8.

Step 2: Obtain the API Key

To send data to ThingSpeak, a unique API key is required, which is used in the main Python code to upload the status of bin to ThingSpeak server. Navigate to 'API Keys' (Fig. 4.6) header under the newly created above channel to get the unique API key as shown in Fig. 4.9.

Step 3: Configuring ESP8266 WiFi Wireless Module

Download Esp8266 WiFi library for Pyboard: Visit the website https://www.tinyosshop.com/wifi-skin-for-pyboard and click on 'Test code' to download the 'pywifi' library (Fig. 4.10a). After extracting the downloaded file, it shows two MicroPython files, i.e. '*main.py*' and '*pywifi.py*' as shown in Fig. 4.10a. Copy '*pywifi.py*' file and open PyCharm IDE → paste it in a PyCharm IDE by saving it with name '*pywifi.py*' in Pyboard.

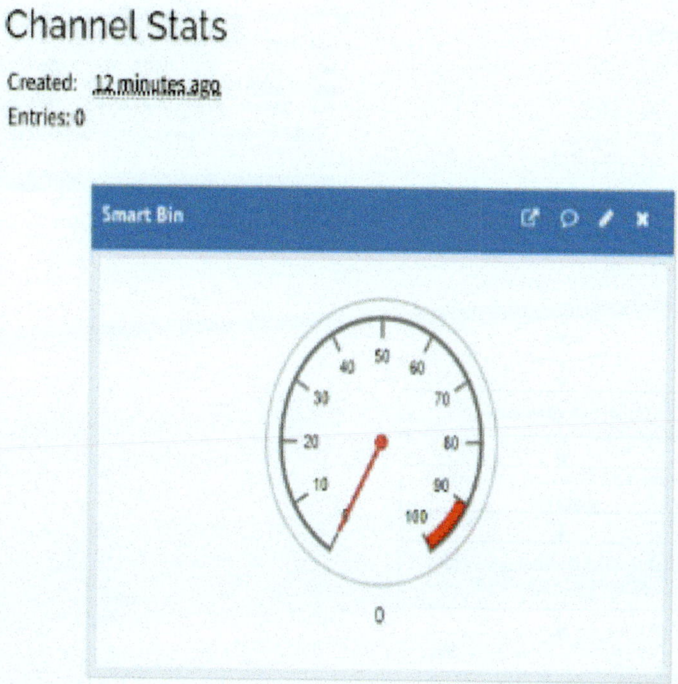

Fig. 4.8 Created widget

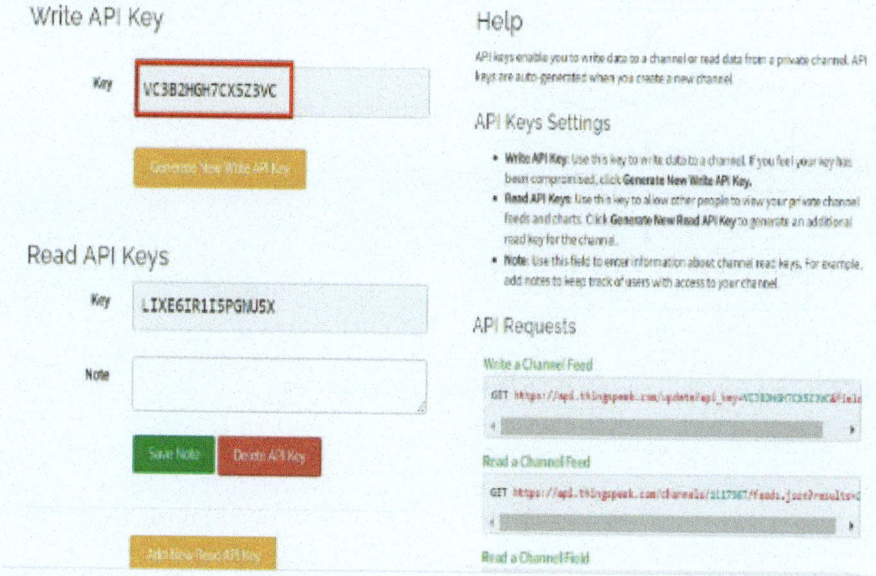

Fig. 4.9 ThingSpeak write key access

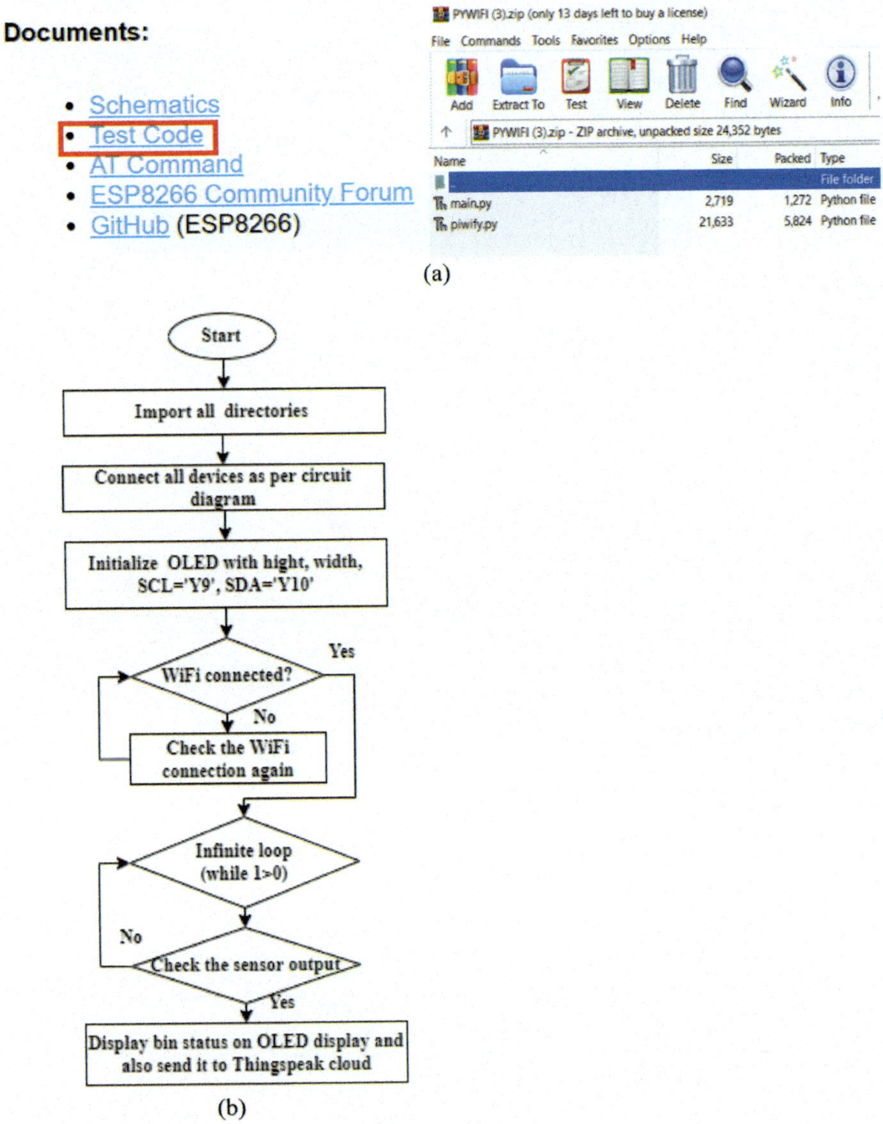

Fig. 4.10 **a** pywifi library details and **b** Flowchart for e-waste bin

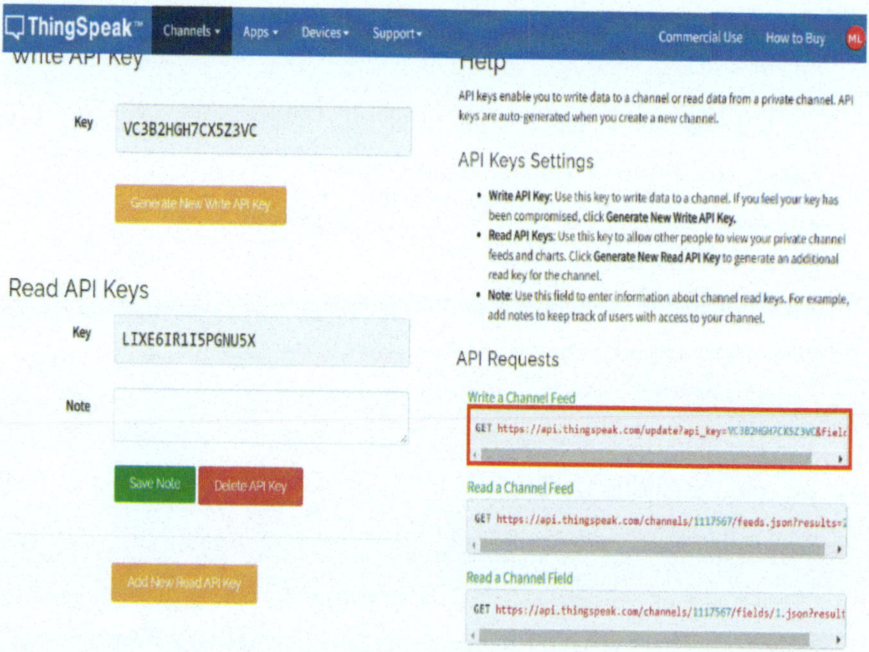

Fig. 4.11 ThingSpeak write channel feed access and channel status

Step 6: complete code (main.py)

Enter the below code in main.py (Fig. 4.11):

```
from pyb import Pin
import pyb
import ssd1306
import machine
from micropython import const
from machine import UART
import pywifi
width = const (128)
height= const (64)
ssd1306_scl= Pin('Y9', Pin.OUT_PP)
ssd1306_sda= Pin('Y10', Pin.OUT_PP)
i2c_ssd1306= machine.I2C(scl=ssd1306_scl, sda=ssd1306_sda)
oled = ssd1306.SSD1306_I2C(width, height, i2c_ssd1306)
```

```
while 1:
    rst_pyb = Pin('X11', Pin.OUT)
    rst_pyb.low()
    pyb.delay(20)
    rst_pyb.high()
    pyb.delay(500)

    Pyboard_wifi = pywifi.ESP8266(1, 115200)

    wifi_mode = 3

    Pyboard_wifi.set_mode(wifi_mode)

    pyb.delay(50)

    Pyboard_wifi.connect(ssid='AndroidAP93E9', psk='uxbm0411')

    pyb.delay(50)
    oled.fill(0)
    oled.text('WiFi Connected', 0, 0)
    oled.show()
    pyb.LED(4).on() #BLUE LED ON

    p_in1 = Pin('X1',Pin.IN, Pin.PULL_UP)
    p_in2 = Pin('X2', Pin.IN, Pin.PULL_UP)
    p_in3 = Pin('X3', Pin.IN, Pin.PULL_UP)
       # The dest_ip for Thingspeak website is 184.106.153.149

    Pyboard_wifi.start_connection(protocol='TCP',dest_ip='184.106.153.149',
                                  dest_port=80, debug=True)
    if p_in1.value() == True and p_in2.value() == True and p_in3.value() == True:
        oled.text('Bin is Empty', 0, 10)
        bin='0'
        oled.show()
    if p_in1.value() == False and p_in2.value() == True and p_in3.value() == True:
        oled.text('Bin is less than Half', 0, 10)
        bin='33'
        oled.show()
    if p_in1.value() == False and p_in2.value() == False and p_in3.value() == True:
        oled.text('Bin is Half', 0, 10)
        bin='66'
        oled.show()
    if p_in1.value() == False and p_in2.value() == False and p_in3.value() == False:
        oled.text('Bin is Full', 0, 10)
        bin='100'
        oled.show()

    #Copy write channel feed from Thingspeak website as shown in figure 4.11

    Pyboard_wifi.send('GET
    https://api.thingspeak.com/ update? api key=HKFRK1JCH5BMOCQ8&field1= '
    + str(bin) + ' HTTP/1.0\r\nHost:192.168.43.176\r\n\r\n', debug=True)

    pyb.delay(1000)
```

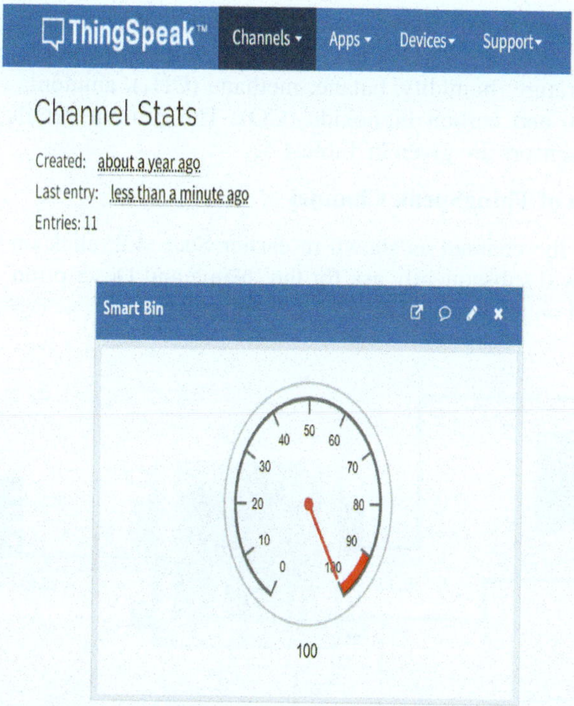

Fig. 4.12 Status of smart bin on ThingSpeak

After entering the above code, one has to run the '*main.py*' program and the status of the bin will be displayed on OLED display as well as on ThingSpeak cloud shown in Fig. 4.12.

4.3 Industrial Environmental Monitoring

Environmental monitoring is essential for protecting both human health and the ecosystem. Industries are the backbone of today's contemporary society, allowing for mass production of commodities to meet the needs of an ever-increasing population. Unfortunately, industrial pollution has a detrimental impact on the planet we live in. Hence, it's our responsibility to give our next generation a greener world for a better life. Governments around the globe have created norms and regulations to monitor the industries and to keep the pollution under check. Routine environmental monitoring data can also be used to validate and compare results about chemical behavior based on laboratory or field investigations (Artiola and Brusseau 2019).

The authors have implemented industrial environmental monitoring systems using IoT (Fig. 4.13). The designed system monitors industrial environment parameters such as temperature, humidity, butane, methane (CH_4), ammonia (NH_3), nitrogen dioxide (NO_2), and carbon monoxide (CO). The pin-out details for interfacing Pyboard with sensors are given in Table 4.4.

Configuration of ThingSpeak Channel

After creating the channel as shown in earlier Sect. 4.2, click on 'New Channel' button which will subsequently ask for the 'Name and Description' of the data you want to upload on this channel as shown in Fig. 4.14a.

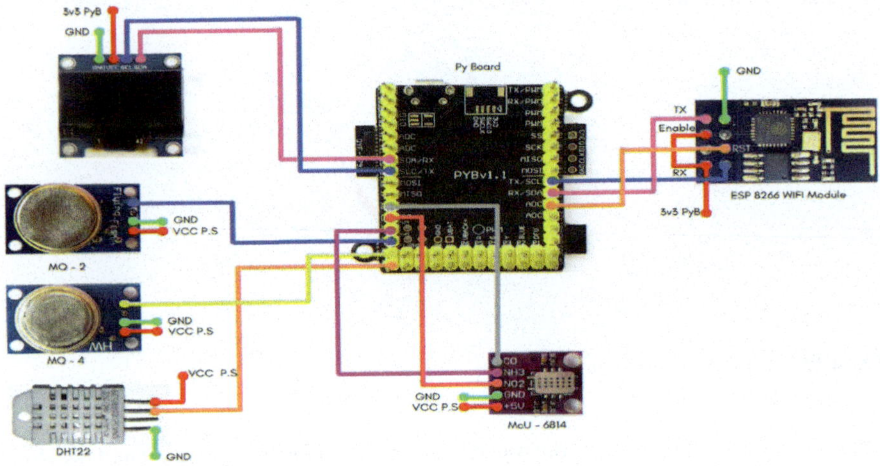

Fig. 4.13 Circuit diagram of industrial environmental monitoring

Table 4.4 Pin connection of Pyboard with various sensors

Pyboard	Sensors	Pin	Uses
X1	DHT22	data	To detect temperature and humidity
X2	MQ4	Analog out	To detect methane gas
X3	MQ2	Analog out	To detect butane
X4	CJMCU 6814	NH_3	To detect NH_3
X5	CJMCU 6814	NO_2	To detect NO_2
X6	CJMCU 6814	CO	To detect CO

In this application, the authors have named the channel 'Industrial Environmental Monitoring' and seven fields are created as shown in Fig. 4.14. After this, click on 'save channel' button to save the details.

Obtain the API Key

To send data to ThingSpeak, a unique API key is required, which is used in '*main.py*' code to upload parameters onto the ThingSpeak server. Navigate to 'API Keys' header under the newly created channel to get the unique API key. After completing these steps, the channel is ready to receive the parameters. Enter the below code in '*main.py*' and save it on Pyboard. The flowchart for industrial environmental monitoring system is shown in Fig. 4.14b.

(a)

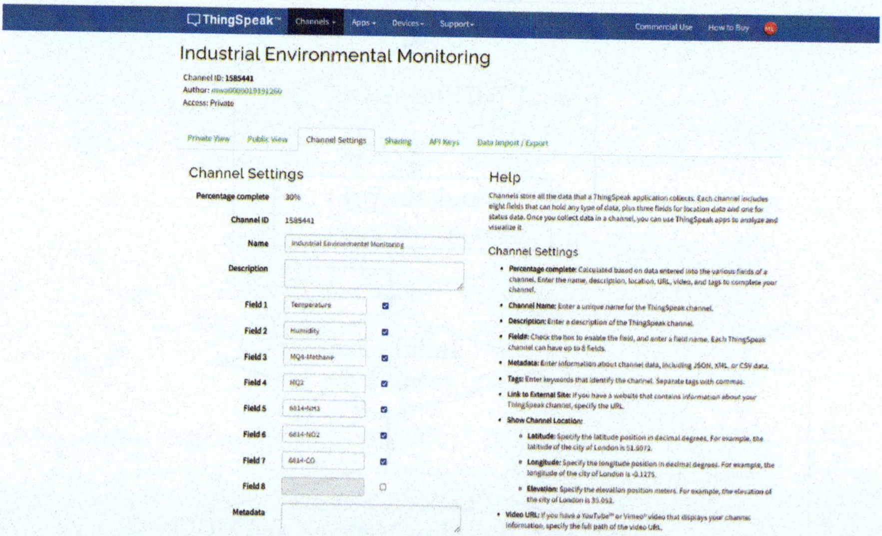

Fig. 4.14 **a** Channel settings and **b** Flowchart for Industrial Environmental Monitoring system

(b)

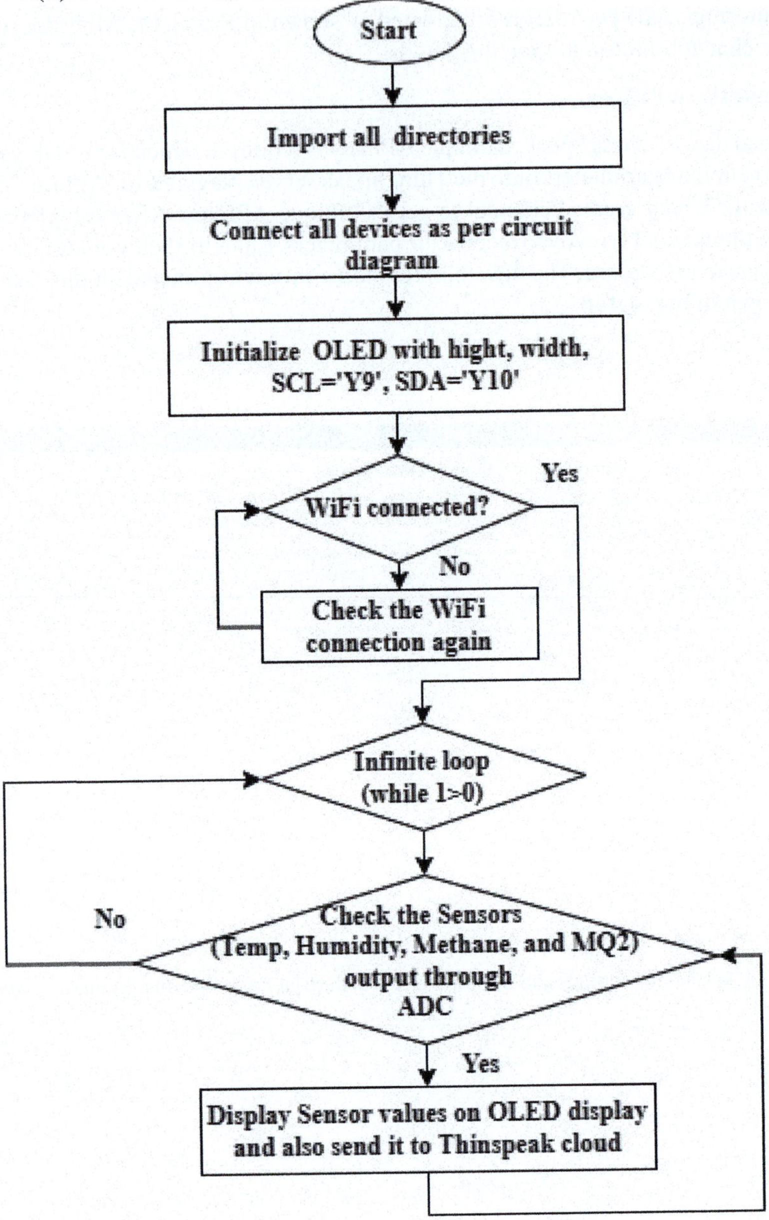

Fig. 4.14 (continued)

Complete code (main.py)

```
from pyb import Pin
import pyb
import ssd1306
import machine
from machine import Pin
from micropython import const
import dht
width = const (128)
height= const (64)
ssd1306_scl= Pin('Y9', Pin.OUT_PP)
ssd1306_sda= Pin('Y10', Pin.OUT_PP)
i2c_ssd1306=machine.I2C(scl=ssd1306_scl, sda=ssd1306_sda)
oled = ssd1306.SSD1306_I2C(width, height, i2c_ssd1306)
import pywifi
temp_hum = dht.DHT22(Pin('X1'))
while 1:
    rst_pyb = Pin('X11', Pin.OUT)
    rst_pyb.low()
    pyb.delay(20)
    rst_pyb.high()
    pyb.delay(500)
    Pyboard_wifi = pywifi.ESP8266(1, 115200)
    wifi_mode = 3
    Pyboard_wifi.set_mode(wifi_mode)
    pyb.delay(50)
    Pyboard_wifi.connect(ssid='AndroidAP93E9', psk='uxbm0411')
    pyb.delay(50)
    pyb.LED(4).on() #BLUE LED ON
    pyb.delay(2000)
    temp_hum.measure()
    temp = temp_hum.temperature()
    hum = temp_hum.humidity()
    MQ4 = pyb.ADC('X2')  # create an analog object for Methane
    Out1 = MQ4.read()  # read an analog value
    MQ2 = pyb.ADC('X3')  # create an analog object for Butane
    Out2 = MQ2.read()  # read an analog value
    NH3 = pyb.ADC('X4')  # create an analog object for Ammonia
    Out3 = NH3.read()  # read an analog value
    NO2 = pyb.ADC('X5')  # create an analog object for Nitrogen Dioxide
    Out4 = MQ2.read()  # read an analog value
    CO = pyb.ADC('X6')  # create an analog object for Carbon Monoxide
    Out5 = CO.read()  # read an analog value'''
    oled.text("Tem: " + str(temp), 0, 10)
    oled.text("Humidity:" + str(hum), 0, 20)
    oled.text("Methane:" + str(Out1), 0, 30)
    oled.text("MQ2:" + str(Out2), 0, 40)
    oled.show()
    oled.fill(0)
    pyb.delay(1000)
    oled.text("NH3:" + str(Out3), 0, 10)
    oled.text("NO2: " + str(Out4), 0, 20)
    oled.text("CO:" + str(Out5), 0, 30)
    oled.show()
    pyb.delay(1000)
    Pyboard_wifi.start_connection(protocol='TCP',dest_ip='184.106.153.149',
    dest_port=80,debug=True)
```

```
Pyboard_wifi.send('GET
https://api.thingspeak.com/update?api_key=7RZ93ORH43F2RYP7&field1=' +
str(temp)+'&field2='+ str(hum) +'&field3='+ str(Out1)+'&field4='+
str(Out2)+'&field5='+ str(Out3)+'&field6='+ str(Out4)+'&field7='+ str(Out5)+'
HTTP/1.0\r\nHost: 192.168.43.176\r\n\r\n', debug=True)
pyb.delay(1000)
```

After entering the above code, one has to execute the '*main.py*' program and the industrial environmental parameters will be displayed on OLED display as well as on ThingSpeak cloud as shown in Fig. 4.15.

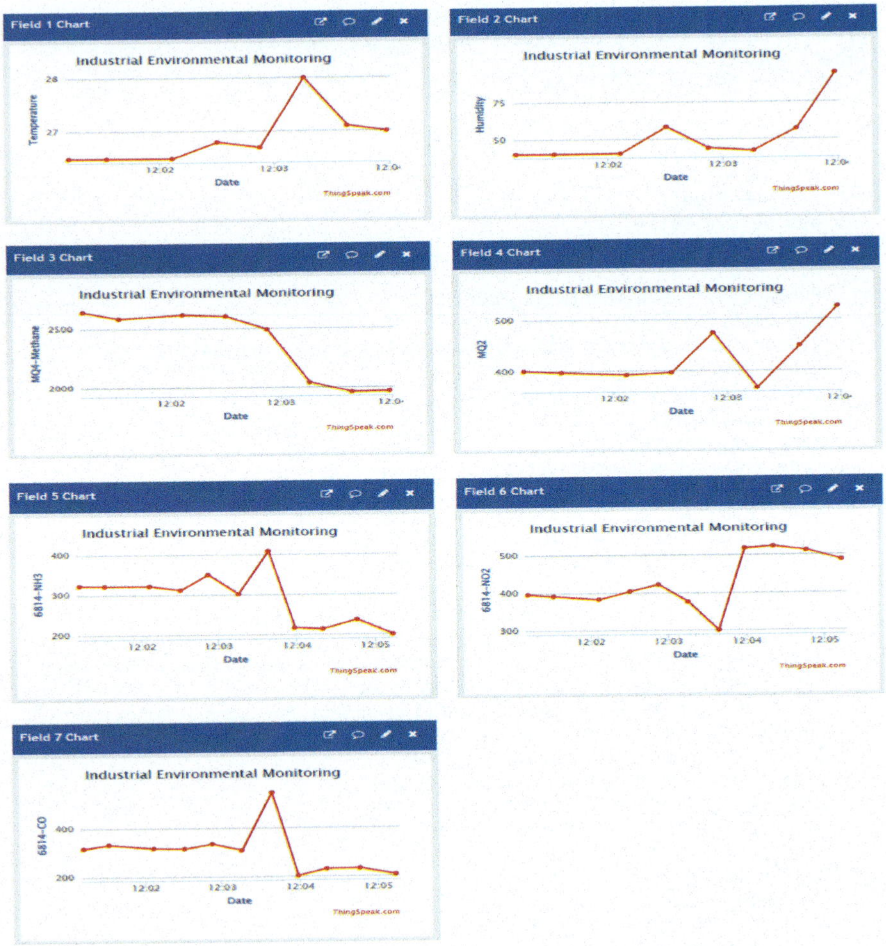

Fig. 4.15 Industrial environmental parameters monitoring system on ThingSpeak

4.4 Greenhouse Monitoring

Flora such as flowers and vegetables are grown in a greenhouse. Greenhouses warm up during the day as sunlight passes through them, heating the plants, soil, and structure. Greenhouses protect crops from a variety of illnesses, notably these are soil-borne and splash onto plants in the rain. The greenhouse effect is a natural phenomenon that is advantageous to humans. To achieve optimal plant growth, the system must be continuously monitored and managed, for example, temperature, moisture, soil humidity, light intensity, and so on (Environmental and Pollution Science 2019). Many farmers fail to profit from greenhouse crops because they are unable to control two critical elements that influence plant development and output. The temperature of the greenhouse should be maintained at a specified level. Crop transpiration and condensation of water vapor on various greenhouse surfaces can all be caused by high humidity. This greenhouse monitoring and management system comes to the rescue in the face of such obstacles. The design and implementation of several sensors for greenhouse environment monitoring, such as humidity, temperature, light condition, and soil moisture, are discussed here.

The authors have implemented greenhouse monitoring systems using IoT. The designed system monitors parameters such as temperature, humidity, water level, and light. The pin-out details for interfacing Pyboard with sensors are given in Table 4.5. The temperature and humidity sensors detect temperature and humidity, the soil moisture sensor detects water level, and the LDR sensor detects light. The detailed interfacing diagram is shown in Fig. 4.16.

Configuration of ThingSpeak Channel

After creating the channel as shown in earlier Sect. 4.2, click on 'New Channel' button which will subsequently ask for the 'Name and Description' of the data you want to upload on this channel as shown in Fig. 4.17.

In this application, the authors have named the channel 'Greenhouse Monitoring' and four fields are created as shown in Fig. 4.17a. After this, click on 'save channel' button to save the details. Figure 4.17b shows the flowchart for greenhouse monitoring system.

Table 4.5 Pin connection of Pyboard with sensors to monitor Greenhouse

Pyboard	OLED	DHT22	LDR	Moisture
Y9	SCL	–	–	–
Y10	SDA	–	–	–
X1	–	Data	–	–
X2	–	–	Output of circuit	–
X3	–	–	–	Aout

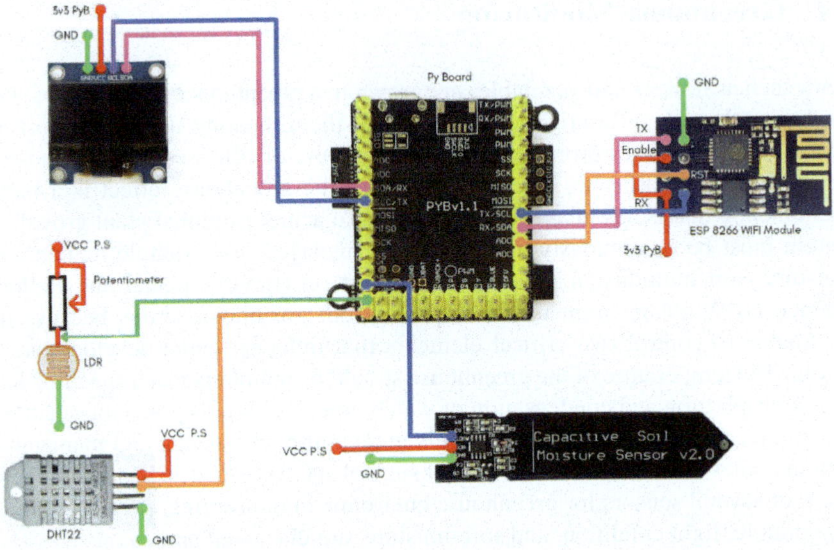

Fig. 4.16 Circuit diagram of Greenhouse monitoring

(a)

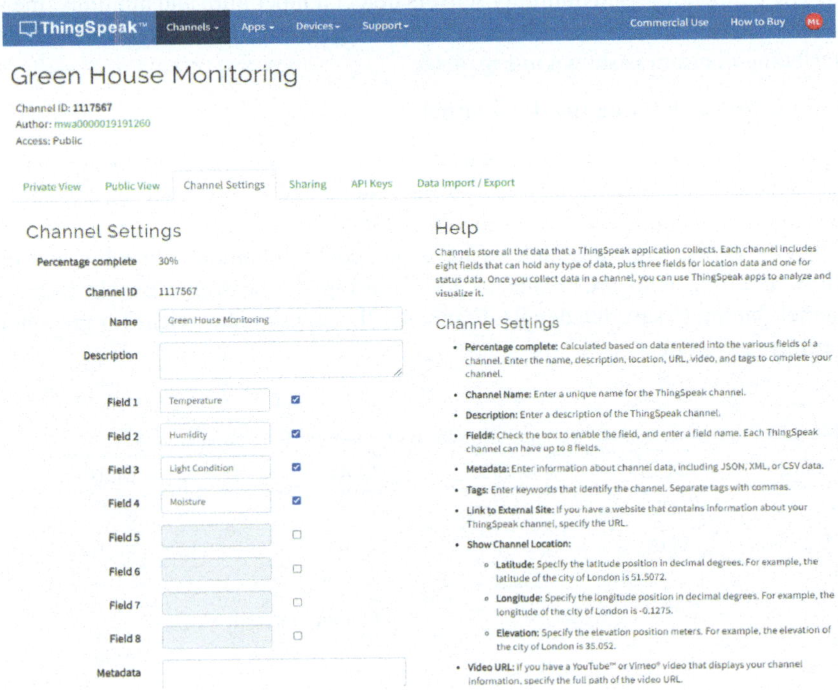

Fig. 4.17 **a** Channel settings and **b** flowchart for greenhouse monitoring system

(b)

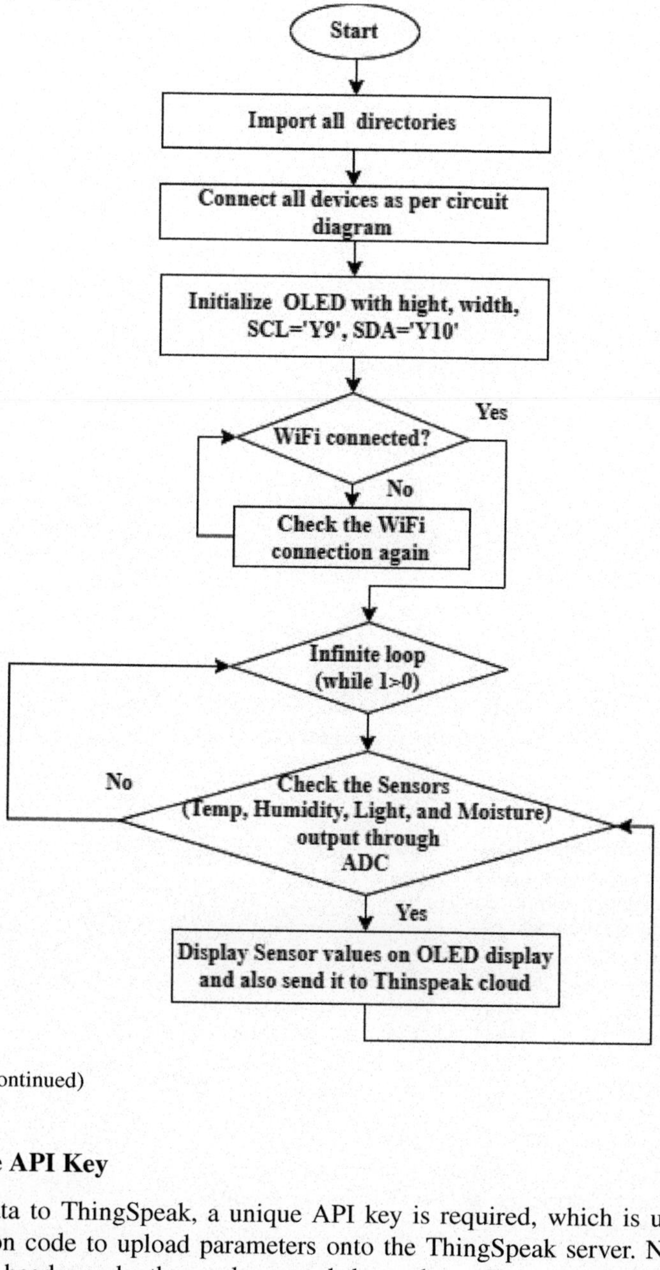

Fig. 4.17 (continued)

Obtain the API Key

To send data to ThingSpeak, a unique API key is required, which is used in the main python code to upload parameters onto the ThingSpeak server. Navigate to 'API Keys' header under the newly created above channel to get the unique API key. After completing these steps, the channel is ready to receive the parameters. Enter the below code in '*main.py*' and save it on Pyboard.

#Complete code (main.py)

```
from pyb import Pin
import pyb
import ssd1306
import machine
from machine import Pin
from micropython import const
import dht
width = const (128)
height= const (64)
ssd1306_scl= Pin('Y9', Pin.OUT_PP)
ssd1306_sda= Pin('Y10', Pin.OUT_PP)
i2c_ssd1306=machine.I2C(scl=ssd1306_scl, sda=ssd1306_sda)
oled = ssd1306.SSD1306_I2C(width, height, i2c_ssd1306)

import pywifi
temp_hum = dht.DHT22(Pin('X1'))
while 1:
    rst_pyb = Pin('X11', Pin.OUT)
    rst_pyb.low()
    pyb.delay(20)
    rst_pyb.high()
    pyb.delay(500)
    Pyboard_wifi = pywifi.ESP8266(1, 115200)
    wifi_mode = 3
    Pyboard_wifi.set_mode(wifi_mode)
    pyb.delay(50)
    Pyboard_wifi.connect(ssid='AndroidAP93E9', psk='uxbm0411')
    pyb.delay(50)
    pyb.LED(4).on() #BLUE LED ON
    pyb.delay(2000)
    temp_hum.measure()
    temp = temp_hum.temperature()
    hum = temp_hum.humidity()
    LDR = pyb.ADC('X2')
    Out1 = LDR.read()  # read an analog value
    Moisture = pyb.ADC('X3')  # create an analog object from a pin
    Out2 = Moisture.read()  # read an analog value
    mois = ((Out2 - 3700) * (100 - 1) / (1300- 3700) +1)

    oled.text("Tem: " + str(temp), 0, 10)
    oled.text("Humidity:" + str(hum), 0, 20)
    oled.text("Light:" + str(Out1), 0, 30)
    oled.text("Moist-sense:" + str(Out2), 0, 40)
    oled.text("Moisture %:" + str(mois), 0, 50)
    oled.show()
    oled.fill(0)
    pyb.delay(1000)
    Pyboard_wifi.start_connection(protocol='TCP',dest_ip='184.106.153.149',
    dest_port=80,debug=True)
```

```
Pyboard_wifi.send('GET
https://api.thingspeak.com/update?api_key=E3SLU3MT69BN8G3V&fiel
d1=' + str(temp)+'&field2='+ str(hum) +'&field3='+ str(Out1)+  '&field4='+
str(mois)+' HTTP/1.0\r\nHost: 192.168.43.176\r\n\r\n', debug=True)

pyb.delay(1000)
```

After entering the above code, one has to execute the '*main.py*' program and the greenhouse monitoring parameters will be displayed on OLED display as well as on ThingSpeak cloud shown in Fig. 4.18.

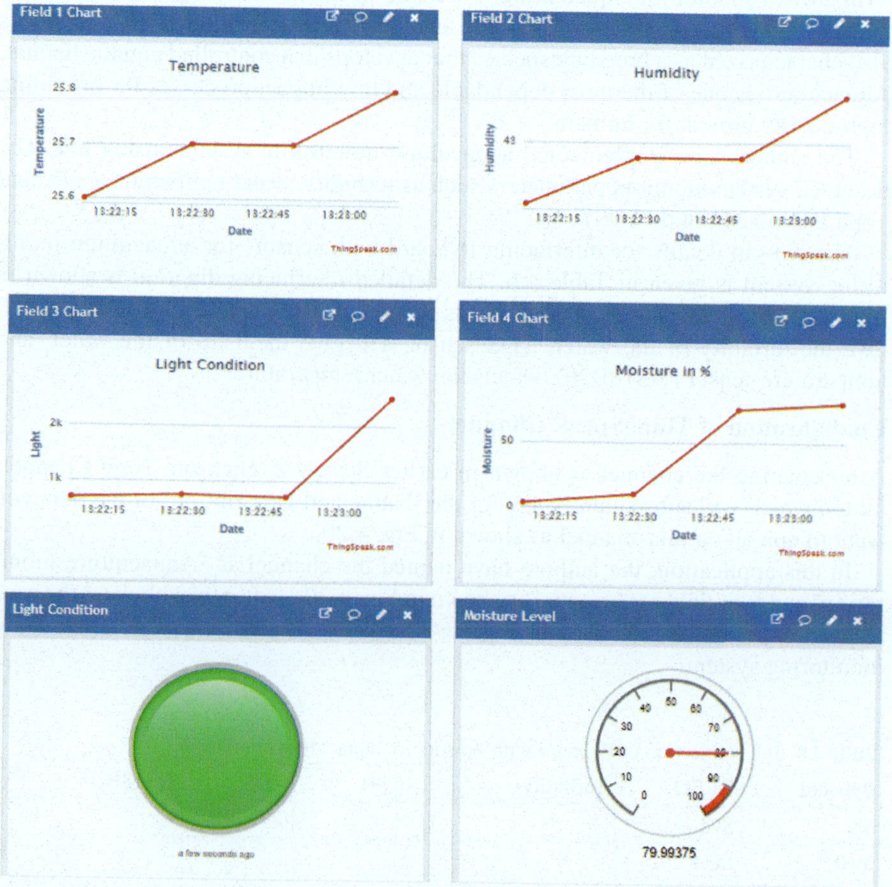

Fig. 4.18 Greenhouse monitoring and control system on ThingSpeak

4.5 Aquaculture Monitoring

Aquaculture is a fast-expanding food production method that has successfully increased fish and shellfish production which helps to feed the world's growing population. Aquaculture is the primary means of survival, being the main source of income for a large strata of society. However, this kind of food production has numerous problems, including rising costs, stricter government regulations, and restricted water supplies. These difficulties have necessitated the development of more complex monitoring and feeding equipment in order to provide tightly controlled and long-term growth conditions. Aquaculture, often known as aqua farming, is the breeding, rearing, and harvesting of fish, seaweed, algae, and a variety of other creatures. It is also characterized as a breeding species that develops in a controlled aquatic habitat. Aquaculture is one of the most dependable and low-impact processes for providing high-quality protein for humans.

The authors have implemented aquaculture monitoring systems using IoT. The designed system monitors parameters such as turbidity, water temperature, pH, and Total Dissolved Solid (TDS).

The pin-out details for interfacing Pyboard with sensors for aquaculture monitoring system is given in Table 4.6. The detailed interfacing diagram is shown in Fig. 4.19. The Grove—PH sensor detects the pH of the water, turbidity sensor will give the turbidity of the water, TDS sensor will give the TDS of the water, and temperature sensor (DS18B20) detects the water temperature.

Configuration of ThingSpeak Channel

After creating the channel as shown in earlier Sect. 4.2, click on 'New Channel' button which will subsequently ask for the 'Name and Description' of the data you want to upload on this channel as shown in Fig. 4.20.

In this application, the authors have named the channel as 'Aquaculture monitoring' and four fields are created as shown in Fig. 4.20a. After this, click on the save channel button to save the details. Figure 4.20b shows the flowchart for aquaculture monitoring system.

Table 4.6 Pin connection of Pyboard with sensors for aquaculture monitoring

Pyboard	OLED	Turbidity	pH	TDS	DS18B20
Y9	SCL	–	–	–	–
Y10	SDA	–	–	–	–
Y11	–	Output of circuit	–	–	–
Y12	–	–	SIG	–	–
X7	–	–	–	Aout	–
X8	–	–	–	–	Sensor output

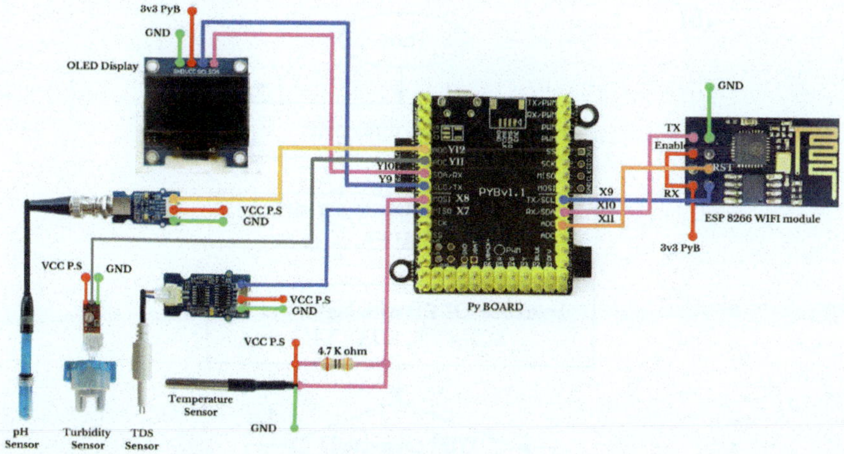

Fig. 4.19 Circuit diagram of aquaculture monitoring

(a)

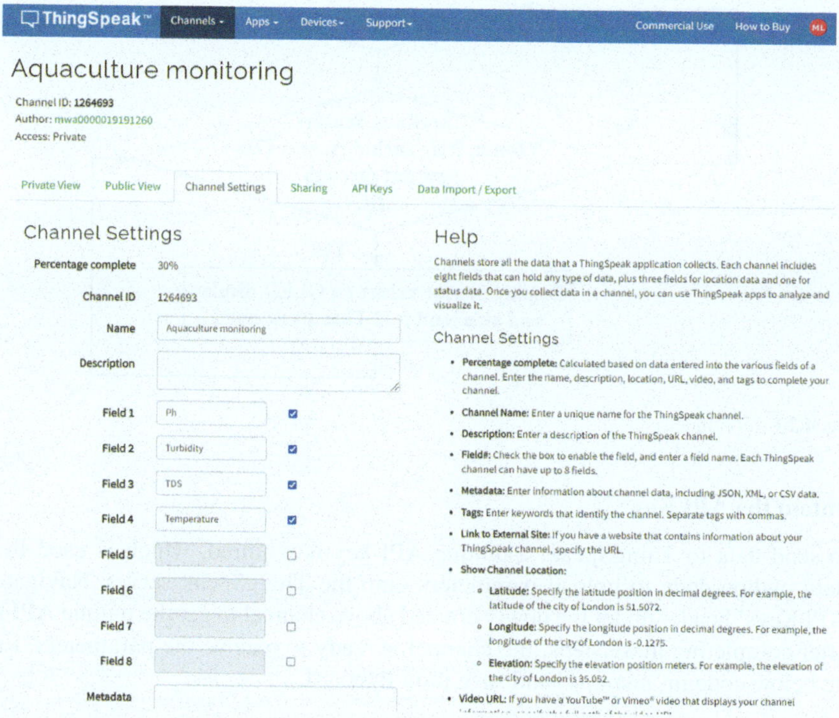

Fig. 4.20 **a** Channel settings. **b** Flowchart for aquaculture monitoring system

(b)

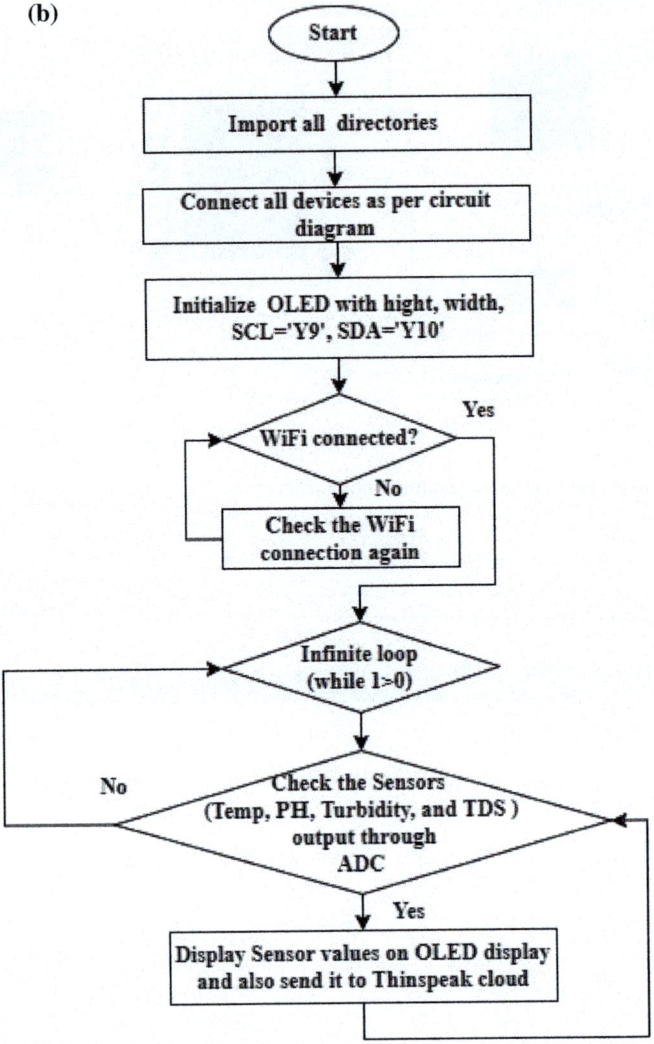

Fig. 4.20 (continued)

Obtain the API Key

To send data to ThingSpeak, a unique API key is required, which is used in the main python code to upload parameters onto the ThingSpeak server. Navigate to 'API Keys' header under the newly created above channel to get the unique API key. After completing these steps, the channel is ready to receive the parameters. Enter the below code in '*main.py*' and save it on Pyboard.

```
#Complete code (main.py)
import time
import ssd1306
import machine
from pyb import Pin
from micropython import const
import pyb
import onewire
import ds18x20

width = const(128)
height = const(64)
ssd1306_scl = Pin('Y9', Pin.OUT_PP)
ssd1306_sda = Pin('Y10', Pin.OUT_PP)
i2c_ssd1306 = machine.I2C(scl=ssd1306_scl, sda=ssd1306_sda)
oled = ssd1306.SSD1306_I2C(width, height, i2c_ssd1306)
import pywifi
while True:
    rst_pyb = Pin('X11', Pin.OUT)
    rst_pyb.low()
    pyb.delay(20)
    rst_pyb.high()
    pyb.delay(500)
    Pyboard_wifi = pywifi.ESP8266(1, 115200)
    wifi_mode = 3
    Pyboard_wifi.set_mode(wifi_mode)
    pyb.delay(50)
    Pyboard_wifi.connect(ssid='AndroidAP93E9', psk='uxbm0411')
    pyb.delay(50)
    pyb.LED(4).on()  # BLUE LED ON
    pyb.delay(2000)
    oled.fill(0)
    adc = pyb.ADC('Y12')    # Ph sensor
    adc1 = pyb.ADC('Y11')    #Turbidity sensor
    val1 = adc1.read()
    adc2 = pyb.ADC('X8')  # TDS sensor
    val2 = adc2.read()
    ds_pin = Pin('X7')      # temperaute sensor
    sensorSum = 0
    for i in (0, 50):
        sensorValue = adc.read()
        sensorSum = sensorSum + sensorValue
    sensorValue = sensorSum / 50
    ph = 7 - 1000 * (sensorValue - 372) * 3.3/59.16/4095
    scaled_value = (val1 - 60) * (0.2 - 5) / (2200 - 60) + 5
    V = val2 * (3.3 / 4095.0)
    tdsValue = (133.42 / V * V * V - 255.86 * V * V + 857.39 * V) * 0.5
    ds_sensor = ds18x20.DS18X20(onewire.OneWire(ds_pin))
    roms = ds_sensor.scan()
    ds_sensor.convert_temp()
    time.sleep(1)
```

```
for rom in roms:
    oled.text("Ph:" + str(ph), 0, 0)
    oled.text("Turbidity:" + str(scaled_value), 0, 10)
    oled.text("TDS:" + str(tdsValue), 0, 20)
    a=ds_sensor.read_temp(rom)
    oled.text("Temp " + str(a), 0, 30)
    oled.show()
    time.sleep(1)

oled.show()

Pyboard_wifi.start_connection(protocol='TCP',dest_ip='184.106.153.149',
dest_port=80,debug=True)

Pyboard_wifi.send('GET
https://api.thingspeak.com/update?api_key=UBY89DB50V2UPGGU&field1=' + str(ph) +
'&field2=' + str(scaled_value) + '&field3=' + str(tdsValue) + '&field4=' + str(a)
+ ' HTTP/1.0\r\nHost: 192.168.43.176\r\n\r\n', debug=True)

pyb.delay(1000)
```

After entering the above code, one has to execute the '*main.py*' program and the greenhouse monitoring parameters will be displayed on OLED display as well as on ThingSpeak cloud shown in Fig. 4.21.

Conclusion:

PyCharm supports various boards including Pyboard. To start with, MicroPython installation and testing steps of Pyboard in PyCharm IDE are given in detail. The

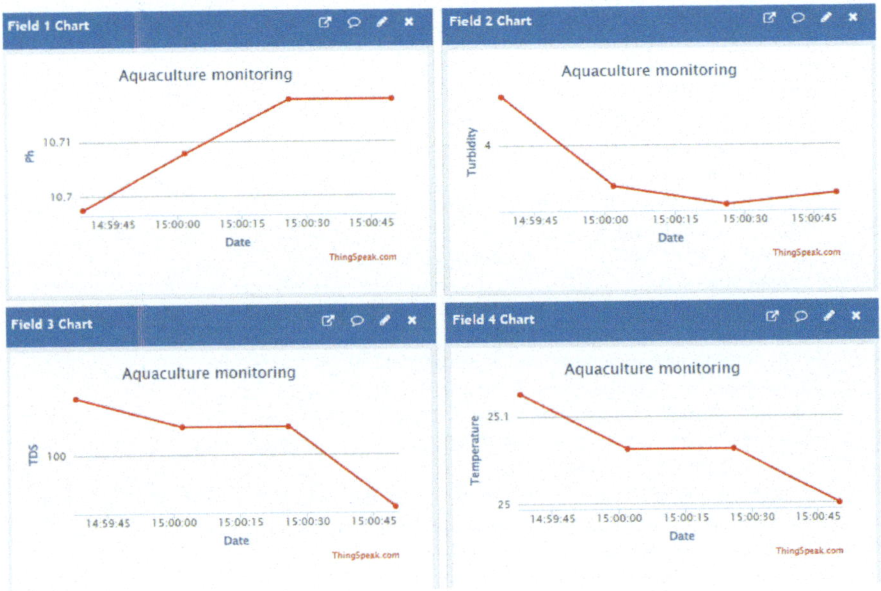

Fig. 4.21 Aquaculture monitoring system on ThingSpeak

various applications, such as, home automation, smart e-waste bin, industrial environmental monitoring, greenhouse monitoring and aquaculture monitoring are implemented to track and control remotely. There is no WiFi support on Pyboards; hence, authors have interfaced external WiFi module. Home automation system is first implemented with switches and appliances interfaced directly to Pyboard. The same application is implemented with ESP32 and controlled remotely with Blynk app. For other applications, the authors have provided detailed interfacing and configuration steps to monitor/control remotely through ThingSpeak cloud.

Exercise

(1) Blink eight LEDs using Pyboard
(2) Design a system to control six appliances with switches connected to ESP32. Also, display the status of each appliance on the OLED screen.
(3) Interface a temperature sensor to ESP32. Using a relay, build a system to turn ON/OFF the fan when temperature rises/decreases of a particular threshold temperature and monitor the temperature.
(4) Assume there are two rooms, Room1 and Room2. Each room has a temperature sensor and a fan respectively. Design a system that can monitor the temperature of each room separately and turn ON/OFF the fan in the respective room if temperature crosses the threshold.
(5) Using turbidity sensor along with ESP32, build a system to monitor cleanliness of water. If the value of turbidity is higher than a set threshold, give out an alarm signaling the need for water replacement/cleaning. The system can then be applied to monitor cleanliness of water in aquariums.
(6) Interface five IR sensors to get the accurate status of the e-waste bin by placing appropriately and displaying the status on ThingSpeak Cloud.
(7) Interface other industrial sensors to monitor Environmental parameters.

References

https://doi.org/10.1016/j.jksuci.2020.08.016.
https://doi.org/10.1016/B978-0-12-814719-1.00010-0.
https://doi.org/10.1016/B978-0-12-386454-3.01041-1.
Artiola JF, Brusseau ML (2019) The role of environmental monitoring in pollution science. In: Brusseau ML, Pepper IL, Gerba CP (eds) Environmental and pollution science, 3rd edn. Academic Press, pp 149–162. https://doi.org/10.1016/B978-0-12-814719-1.00010-0
Covaci A (2014) Environmental fate and behavior. In: Wexler P (ed) Encyclopedia of toxicology, 3rd edn. Academic Press, pp 372–374. https://doi.org/10.1016/B978-0-12-386454-3.01041-1
Environmental and Pollution Science (Third Edition) (2019) Academic Press, pp 149–162
Madhesh A, Kowsigan M, Khishore R, Balasubramanie P (2020) IoT based home automation and security system. Int J Adv Sci Technol 29(9s):2733–2739. http://sersc.org/journals/index.php/IJAST/article/view/15439
Majeed R, Abdullah NA, Ashraf I, Zikria YB, Mushtaq MF, Umer M (2020) An intelligent, secure, and smart home automation system. Sci Program. ID 4579291, 14 p. https://doi.org/10.1155/2020/4579291

Nasir OA, Mumtazah S (2020) IOT-based monitoring of aquaculture system. Matter: Int J Sci Technol 6(1):113–137. https://doi.org/10.20319/mijst.2020.61.113137

Rahman MW, Islam R, Hasan A, Bithi NI, Hasan MM, Rahman MM (2020) Intelligent waste management system using deep learning with IoT. J King Saud Univ-Comput Inf Sci. https://doi.org/10.1016/j.jksuci.2020.08.016

Stolojescu-Crisan C, Crisan C, Butunoi B-P (2021) An IoT-based smart home automation system. Sensors 21:3784. https://doi.org/10.3390/s21113784

Sujin JS et al (2021) IOT based greenhouse monitoring and controlling system. J Phys: Conf Ser 1916 012062

Zeb A, Ali Q, Saleem MQ, Awan KM, Alowayr AS, Uddin J, Iqbal S, Bashir F (2019) A proposed IoT-enabled smart waste bin management system and efficient route selection. J Comput Netw Commun. ID 7043674, 9 p. https://doi.org/10.1155/2019/7043674

Chapter 5
FoG and Cloud Computing with Jetson Nano

Abstract FoG and cloud paradigms are helping the IoT devices to offload the computing in order to reduce power consumption and increase battery life. The Jetson Nano SBC is an exciting platform which can act as a FoG node and as an intermediate layer between the cloud and IoT devices. The Jetson Nano module is a small Artificial Intelligence (AI) computer that has the performance and power efficiency needed to run modern AI workloads, multiple neural networks in parallel and process data from several sensors simultaneously. The ongoing chapter introduces the concepts of FoG and Cloud computing along with its role in IoT. At the end, this chapter covers implementation of FoG and Cloud computing.

Keywords FoG · Cloud · Patient monitoring · Home security · Fire alert system · Home safety lock · Surveillance

5.1 Introduction to FoG Computing

Cloud computing has gained widespread acceptance due to the services such as data storage, computing, etc. offered by it at a very affordable cost. To achieve this, centralized data centers are offered by big players. More businesses are able to adopt cloud computing for their operations due to the ready adaptability and reduced cost of cloud services (Chiang and Zhang 2016). This offloads a lot of workload and in turn increases productivity (Khan et al. 2017). To keep up with the market needs, cloud computing was able to evolve and offers better performance, security, and reliability. Today, cloud is able to offer various models for services such as Infrastructure as a Service (IaaS), Platform as a Service (PaaS), and Software as a Service (SaaS).

The world has seen an unprecedented rise in devices connected to the Internet over the decade. According to Cisco system inc, there are around 50 billion connected devices (Evans 2011). Naturally, there is high demand for quality cloud services and IoT devices. Since the overall operation of cloud data servers is centralized, it proves to be a hindrance for IoT devices spread over a large geographic area. Problems such as network congestion and high latency are common occurrences in such scenarios. FoG computing is an extension of the existing cloud platform which

J. S. Parab et al., *Python Programming Recipes for IoT Applications*, Transactions on Computer Systems and Networks, https://doi.org/10.1007/978-981-19-9466-1_5

was put forward by Cisco to resolve the issue faced and elaborated on above. Cisco was the first player to formally put forward the concept of FoG computing (Tang et al. 2015). An additional layer called the FoG layer is introduced in between the cloud data centers and the end-users. In principle, this layer is physically much closer to the end-user or the IoT nodes. It helps to achieve the decentralization of the data centers.

FoG computing together with cloud computing and IoT devices (End devices) forms three-tier architecture as shown in Fig. 5.1 (Taneja and Davy 2016). The bottom Tier 1 consists of the End /IoT devices which are the generators of the data. The IoT devices are equipped with sensors that generate the raw data. Tier 2 is the FoG computing layer. This computing layer generally consists of devices such as gateways, routers, switches, etc. which can provide processing and storage of information. Hence, this layer is sometimes referred to as FoG intelligence. If the devices which form the part of the FoG computing layer wishes to send data to the cloud they can do so on a periodic basis. In FoG computing architecture, not every data packet is redirected to the cloud, instead all real-time analysis and latency sensitive applications have a dependency to run from the FoG layer itself. Tier 3 is named as cloud computing layer, also referred to as the cloud intelligence, which leverages modern infrastructure such as data centers to provide enormous storing and processing capabilities to the lower layers.

Dastjerdi et al. defined FoG computing as a distributed computing environment extending the services of cloud computing to one-hop distance from the user (Dastjerdi et al. 2016). FoG network consists of multiple FoG nodes, which communicates with one another and also in sync with the cloud data center. FoG computing and cloud computing work together to enhance the quality and performance of any system. In this type of computing, one utilizes a network of FoG nodes which are placed at just one leap from the end-user or IoT device. This allows the local data processing on the FoG node rather than on the cloud data center which may be placed geographically at a distant location. In addition to data processing, FoG node may also provide

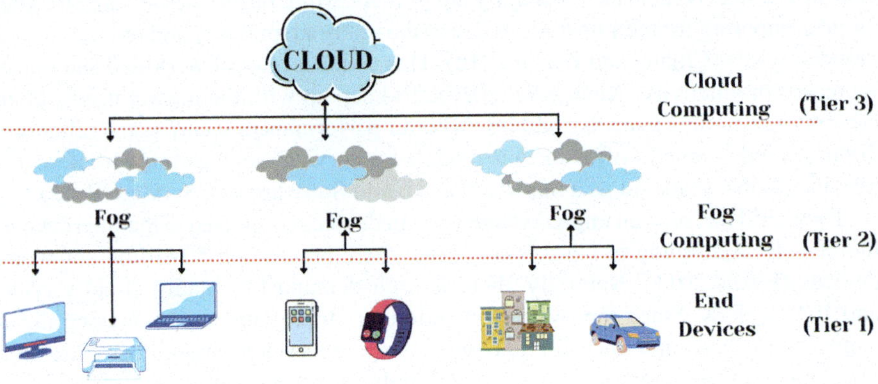

Fig. 5.1 Three-tier architecture

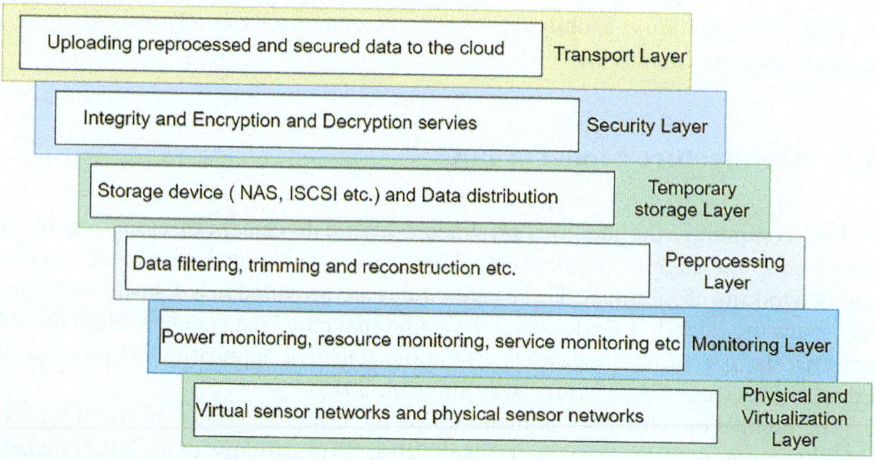

Fig. 5.2 Six-layer FoG architecture

networking between the end-users (edges) and the cloud data centers. The features like communication, mobility, and computing offered by FoG node aim to provide low latency.

FoG Computing Essential Characteristics

This section briefly discusses the essential characteristics of FoG computing. If a device possesses storage, computing capacity, and network connectivity, it can operate as a FoG node. They can be deployed anywhere within a network. Examples of such devices are routers, switches, embedded servers and industrial controllers, and cameras for video surveillance (Atlam et al. 2018). FoG Computing in general should have the below characteristics:

- *Location awareness and low latency*: FoG computing features location awareness and can be set up in any location. As FoG nodes are closer to the end devices, FoG computing gives low latency operation (Atlam et al. 2018).
- *Scalability*: One of the main features of FoG computing is distributed computing and storage networks which work as end devices/sensor networks. Additionally, FoG computing supports the ready addition of new end devices/sensor networks (Atlam et al. 2018).
- *Geographical distribution*: Services and applications given by FoG nodes are distributed and can be deployed anywhere, in contrast to the services of centralized data servers.
- *Heterogeneity and Interoperability*: FoG nodes or end devices are designed by different manufacturers but still possess the ability to work on different platforms and across different service providers (Atlam et al. 2018).
- *Real-time interactions*: FoG computing gives a real-time interaction between FoG nodes rather than the batch processing model provided by the cloud (Atlam et al. 2018).

- *Support for mobility:* Mobile devices can be connected directly in FoG applications (Atlam et al. 2018).

5.2 Architecture Model of FoG

In FoG computing, the facilities provided by the data centers of cloud computing are taken to the edge of the network. FoG nodes offer limited computing, storing, and networking capabilities. These capabilities are provided in a distributed manner that spans the Edge/IoT devices and the traditional cloud data centers. Here the sole objective is to give minimal latency for time-sensitive application running on the Edge/IoT devices (Atlam et al. 2018; Shi et al. 2015).

The architecture for FoG is made up of six layers as shown in the figure 5.2 (Mukherjee et al. 2018; Aazam and Huh 2014, 2015; Muntjir et al. 2017) namely, physical and virtualization, monitoring, pre-processing, temporary storage, security, and transport layer.

The lowest layer is the physical and virtualization layer and is made up of virtual sensor networks along with physical and wireless sensor networks. In this layer, myriad types of sensors are geographically distributed for environment sensing which generates raw data. This raw data is passed to the next layers via gateways for further processing and filtering (Liu et al. 2017).

The next layer is the monitoring layer which monitors the resource utilization and the availability of FoG nodes along with the other network elements. This layer is also responsible for monitoring services, performance, power consumtion, and tracking the status of applications running on FoG nodes (Mukherjee et al. 2018; Aazam and Huh 2015).

Preprocessing and temporary storage layer handles the tasks of data management. The data forwarded by the lower layers are analyzed, trimmed, and filtered to extract meaningful information. This preprocessed data in the below layer is then required to be temporarily stored in the next layer (Aazam and Huh 2015; Muntjir et al. 2017).

The encryption/decryption of the data on the FoG nodes is handled at the security layer. Additional integrity mechanism is also applied to the data to protect the data from being tampered.

The topmost layer of the architecture is the transport layer where the data is uploaded to the cloud to generate more useful services (Aazam and Huh 2015; Muntjir et al. 2017). As the FoG nodes are constrained by limited resources, the protocols used for communication on the FoG nodes have to be lightweight, efficient, and customizable (Aazam et al. 2014; Marques et al. 2017).

5.3 Introduction to Cloud Computing

Prior to the introduction of cloud computing, if there is a need of computational facility, one has to invest in the hardware, software, networking, and storage requirements for such an endeavor. The cost includes the upfront cost of the real estate to house the hardware, the maintenance, and the operational cost. This becomes an enormous cost for an individual or enterprise needing a large computing facility (Chandrasekaran 2015). With the cloud, it is possible to use the computing facility of a provider as and when it is needed. Hence, cloud computing can be termed as utilizing the computing infrastructure made available by a service provider, to the extent needed, and paying only for the service consumed.

What is Cloud Computing?

The popular definitions among the experts on cloud computing were put forward by the National Institute of Standards and Technology (NIST). It states that "Cloud computing is a model for enabling ubiquitous, convenient, on-demand network access to a shared pool of configurable computing resources (e.g. networks, servers, storage, applications, and services) that can be rapidly provisioned and released with minimal management effort or service provider interaction" (Sunyaev 2020).

Cloud computing Essential Characteristics:

A cloud service should have below essential characteristics such as on-demand self-service, ubiquitous access, multi-tenancy, location independence, rapid elasticity, and metering.

On-demand self-service: A cloud service should allow a user/consumer to independently avail computing resources, such as network storage and server time, without human intervention (Mell and Grance 2011).

Ubiquitous access: The services are disseminated using a broadband network prominently using the Internet. The services can be availed on various devices such as smart phones, tablets, and workstations using standard communication interfaces. This enables the users to avail any cloud service from any device or platform at any given time (Mell and Grance 2011; Iyer and Henderson 2010).

Resource pooling and Multi-tenancy: Computing resource pooling is done at the providers end in order to service users/consumers using a multi-tenant architecture. Sharing computing resources is part of what could make cloud computing economically beneficial (Mell and Grance 2011; Arasaratnam 2011).

Location Independence: A user/consumer wanting to use cloud service is given a sense of location independence, where the user has no knowledge nor control of the physical location of the computing resources. The user may be provided a control of choosing at a higher level of abstraction, such as a choice of country, state, or data center (Mell and Grance 2011).

Rapid Elasticity: A sense of elasticity is provided to the user where the resource can be elastically commissioned and decommissioned to keep pace with the current

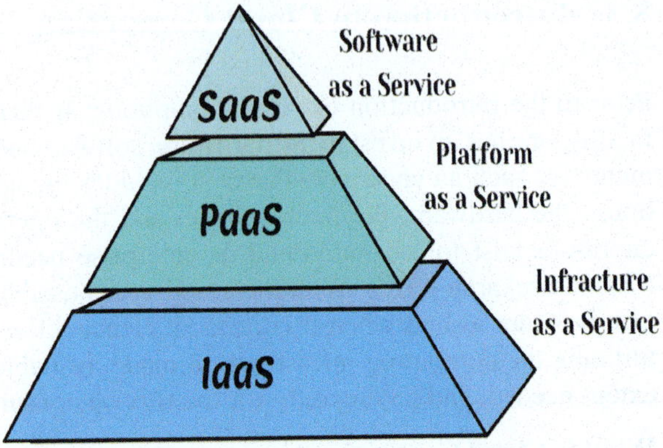

Fig. 5.3 Cloud service pyramid

resources needs. The rapid elasticity gives the users/customers an impression of the availability of unlimited resources.

Metering service: A metering capability is incorporated by the service provider in order to automatically control and optimize the resource utilization. The metering is done for the resources such as bandwidth, storage, processing, etc. This provides transparency to both the consumer and service provider.

Cloud Computing Service Models

The services provided by cloud computing have evolved into three typical service models, namely, (1) Infrastructure as a Service, (2) Platform as a Service, and (3) Software as a Service as shown in Fig. 5.3. One finds a hierarchical organization of these top three models based on the level of abstraction of the capabilities made available by individual layers (Kumar and DuPree 2011).

Infrastructure as a Service (IaaS): In the IaaS scheme, a consumer acquires processing, storage, networking, and other resources from the service provider. Here, the consumer is free to utilize the acquired resources at his discretion by running or deploying arbitrary software of his choice which can include applications and operating systems. In IaaS, the service provider gives to the consumer a large number of virtual resources by dividing a very large physical resource of the infrastructure (). The most popular IaaS service providers in the market are IBM, Microsoft, Rackspace, NTT, Oracle, Fujitsu, and Amazon (Marston et al. 2011).

Platform as a Service (PaaS): In PaaS cloud model, a user/consumer develops applications to run on the cloud infrastructure. The applications are developed using programming languages, services, tools, and libraries supported by the provider (Mell and Grance 2011). The user is not given the control or management of the cloud infrastructure, but given the control of configuration settings for the application hosting environment and applications. A PaaS provides a cloud developer freedom to design, build, test and deploy custom applications (). Leading providers offering PaaS are Amazon, Microsoft, Alibaba, Google, IBM, and Rackspace (Voorsluys et al. 2011).

Software as a Service (SaaS): As the name suggests the user/consumer uses the applications provided by the cloud service provider. These applications are running on the cloud infrastructure. The applications are accessed using a client device running a thin client interface such as a browser or a program interface. The management and control of the cloud infrastructure are not done by the user/consumer, while a limited user specific application configuration setting is controlled by the user (Mell and Grance 2011). Therefore, user/consumer enjoys abstraction from all the finer details of the current applications running behind the scenes (Arasaratnam 2011;). Leading SaaS providers are Salesforce, Microsoft, SAP, Oracle, Adobe Systems, and IBM.

Deployment Models

There are four different deployment models such as Private, Public, Community, and Hybrid cloud.

Private cloud: The infrastructure is exclusively reserved for use by the single organizations who's services are given to a large number of consumers, namely, business units. The cloud infrastructure may be housed on the premises of the organization or off premises (Mell and Grance 2011).

Community cloud: The infrastructure is exclusively reserved for a particular community of users from organizations that have common concerns. The cloud infrastructure may be housed on the premises of the organization or off premises (Mell and Grance 2011).

Public cloud: The infrastructure is not reserved for specific users and is open to the general public.

Hybrid cloud: Here, two or more distinct cloud deployment models such as community, public, or private are combined together. However, the individual entities remain unique.

5.4 Cloud Computing Architecture

The cloud architecture is split into two parts, namely, frontend and backend. Fig. 5.4 depicts an internal architectural view of cloud computing.

Cloud computing architecture consists of client infrastructure, application, service, runtime, storage, infrastructure, administration, and security (ITCandor 2018).

Frontend: The cloud architecture's frontend refers to the cloud's client infrastructure, which includes all user interfaces and applications.

Backend: Backend refers to the cloud itself which is used by the service provider. It controls resources and implements security measures. It also contains massive storage, virtual applications, virtual computers, traffic management techniques, deployment models, and so on.

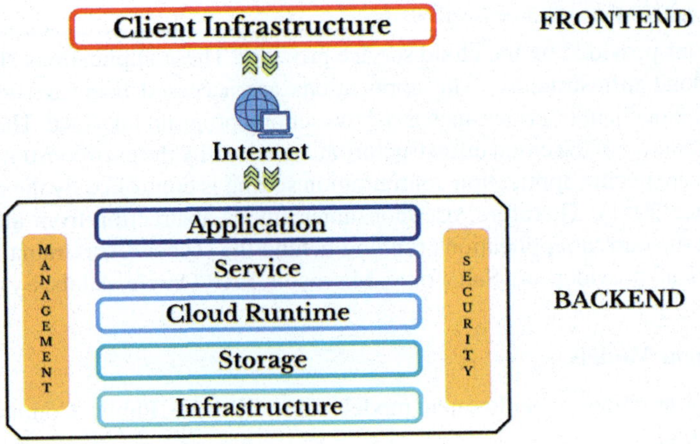

Fig. 5.4 Internal architectural of Cloud

- The Application Layer consists of a software platform that is accessed by the client and offers services in the backend based on the client's needs.
- The Service Layer includes three types of cloud-based services: SaaS, PaaS, and IaaS, and it offers selectable services based on customer requests.
- The Cloud Runtime Layer is in-charge of providing the virtual machine with an execution and runtime platform/environment.
- Storage Layer handles the management and scalability of storage services provided to clients.
- The Infrastructure Layer refers to cloud software and hardware such as servers, storage, network equipment, and so on.
- The Management Layer manages all the backend components.
- Security Layer is responsible to implement different security mechanisms in the backend.
- For interaction and communication, the Internet Layer serves as a bridge between the frontend and the backend.

There are various advantages of Cloud Computing Architecture, such as a simpler overall computing system, improved data processing needs, more security, improved disaster recovery, lower operational costs, and so on.

5.5 Role of FoG and Cloud Computing in IoT

The advent of IoT has seen smart devices permeating every home, vehicle, and workplace with connectivity to the Internet. Due to technological advancements, the things which surround us are potentially getting connected to the Internet and hence there is a huge stress on the current Internet and cloud infrastructure in place.

The conventional approach is to utilize a centralized cloud for processing which is localized at one site. But due to the proliferation of IoT devices, the sheer amount of data generated by these devices is increasing day by day. The existing cloud infrastructure will be severely strained by this massive explosion of data [29].

FoG computing allows computing to happen via IoT devices and only pushes relevant data to the cloud. The FoG brings the cloud closer to the objects that generate and act on IoT data. The FoG nodes can be deployed anywhere with a network between cloud and IoT devices. They analyze the most time-sensitive data at the network edge instead of sending vast amounts of IoT data to the cloud. Only required data can be sent to the cloud for historical analysis and longer term storage. It is very important to point out that FoG does not eliminate the cloud but complements it to improve the efficiency (Schneider and Sunyaev 2016).

5.6 Examples of FoG and Cloud Computing

The FoG and cloud computing has become ubiquitous in the modern computing scenario. On a regular basis, FoG and cloud service providers are adding exciting features to the existing platforms. Here, authors have brought out the salient features of both the paradigms especially relating to IoT with various applications. Under cloud computing, authors have implemented a Patient Monitoring system to monitor Oxygen Saturation, Pulse, and Body Temperature. FoG computing is leveraged for Home Security in which Home Surveillance, Home Safety Lock, and Fire Alert systems are implemented.

5.6.1 Patient Monitoring system with Cloud

In modern society, the incidence of chronic diseases has increased due to different risk factors such as dietary habits, physical inactivity, alcohol consumption, etc. It is said that in the next 10 years, deaths from chronic diseases will increase by 17%. If these diseases are not monitored and treated early, they can lead to several complications and even pose threat to life. Therefore, monitoring of health parameters is of utmost importance (Schneider and Sunyaev 2016).

This application demonstrates the Patient Monitoring system. Here, various health parameters such as Oxygen Saturation (SpO2), Pulse Rate, and Body Temperature are monitored using appropriate sensors. The main heart of this system is Jetson Nano SBC, which is interfaced with these sensors, as shown in Fig. 5.5. The Oxygen Saturation and Pulse Rate are obtained from the MAX30102 sensor which is interfaced with the Jetson nano SBC using I2C. The Body Temperature is obtained using the MLX90614 IR Temperature Sensor which is also interfaced using I2C. The health parameters are updated to ThingSpeak Cloud for monitoring from any geographic location. ThingSpeak is an open IoT platform for monitoring patients data online. Table 5.1 tabulates the pin connection of MAX30102 and MXL90614 with Jetson Nano.

MAX30102 is an integrated pulse oximetry and heart-rate sensor. It integrates two LEDs (IR and Red), a photodetector, optimized optics, and low-noise analog signal processing to detect pulse oximetry and heart-rate signals. It is fully configurable through software registers, and the digital output data is stored in a 32-deep FIFO within the device. It also has an ambient light cancellation (ALC), 18-bit sigma delta ADC, and discrete time filter. It has an ultra-low-power operation which makes it ideal for battery operated systems. MAX30102 operates within a supply range of 1.7 to 2 V. The same sensor can be used for various applications such as fitness assistant, wearable devices, etc.

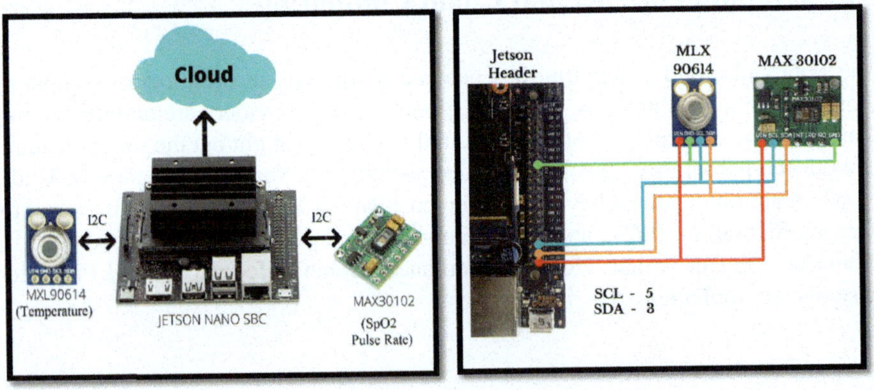

Fig. 5.5 Conceptual and pin diagram of Patient Monitoring system

Table 5.1 Pin connections for patient monitoring

Jetson Nano SBC	MAX30102	MXL90614
3.3 V	VIN	VIN
GND	GND	GND
PIN 3 (I2C 1 SDA)	SDA	SDA
PIN 5 (I2C 1 SCL)	SCL	SCL

MLX90614 sensor is manufactured by Melexis Microelectronics Integrated systems. It works on the principle of infrared thermopile sensor for temperature measurement. These sensors consist of two units embedded internally to give the temperature output. The first unit is the sensing unit which has an infrared detector, followed by the second unit which performs the computation of the data using Digital Signal Processing (DSP). This sensor works on Stefan-Boltzmann law which explains power radiated by a black body in terms of its temperature. MLX90614 sensor converts the computational value into 17-bit ADC and that can be accessed using the I2C protocol. This sensor measures the ambient temperature as well as object temperature with the resolution of 0.02 °C and can be operated with a supply ranging from 3.6 V to 5 V.

Implementation and Configuration steps:

Step 1: Before using I2C on Jetson Nano SBC, one need to install the relevant libraries by executing the following command on Jetson Nano terminals.

```
sudo apt-get install python-setuptools
sudo apt-get install -y i2c-tools
```

There are two I2C ports available on the J41 Header of the Jetson Nano SBC. Here, authors have used the I2C-1 for both the sensors as shown in Fig. 5.5. Pin 3 and Pin 5 of J41 header is SDA and SCL of I2C-1, respectively. One can test if the I2C device is properly connected by using the below command.

```
sudo i2cdetect -y -r 1
```

Step 2: Configuration of Library for MAX30102.

Create a folder named "Patient Monitoring system" → Download the library for MAX30102 from the github link: https://github.com/doug-burrell/max30102 → Extract the downloaded file and place it in the above created folder. A few Python libraries, namely, 'smbus' and 'numpy', are required. Use "apt" to install the these libraries.

```
sudo apt install python- smbu s
sudo apt install python-numpy
```

There are some changes to be done in the file '*heartrate_monitor.py*' of the downloaded library for efficient access to the parameters. The authors have introduced two variables, namely, '*self.print_bpm*' and '*self.print_spo2*' with an initial value as zero inside the class '*HeartRateMonitor (object)*'. These variables are updated later in the same class. The rest of the library files specific to MAX30102 are kept intact. To avoid confusion in the readers mind, authors recommend to adopt the below code.

```
# heartrate_monitor.py
from max30102 import MAX30102
import hrcalc
import threading
import time
import numpy as np

class HeartRateMonitor(object):
    """
    A class that encapsulates the max30102 device into a thread
    """
    LOOP_TIME = 0.01

    def __init__(self, print_raw=False, print_result=False):
        self.bpm = 0
        self.print_spo2 = 0
        self.print_bpm = 0
        if print_raw is True:
            print('IR, Red')
        self.print_raw = print_raw
        self.print_result = print_result

    def run_sensor(self):
        sensor = MAX30102()
        ir_data = []
        red_data = []
        bpms = []

        # run until told to stop
        while not self._thread.stopped:
                # check if any data is available
                num_bytes = sensor.get_data_present()
                if num_bytes> 0:
                        # grab all the data and stash it into arrays
                while num_bytes> 0:
                        red, ir = sensor.read_fifo()
                        num_bytes -= 1
                        ir_data.append(ir)
                        red_data.append(red)
                        if self.print_raw:
                                print("{0}, {1}".format(ir, red))

                while len(ir_data) > 100:
                        ir_data.pop(0)
                        red_data.pop(0)

                if len(ir_data) == 100:
                        bpm, valid_bpm, spo2, valid_spo2 = hrcalc.calc_hr_and_spo2(ir_data,
                                                                                  red_data)
                        if valid_bpm:
                                bpms.append(bpm)
                                while len(bpms) > 4:
                                        bpms.pop(0)
```

```
                    self.bpm = np.mean(bpms)
                    if (np.mean(ir_data) < 50000 and np.mean(red_data) < 50000):
                            self.bpm = 0
                            if self.print_result:
                                    print("Finger not detected")
                    if self.print_result:
                            print("BPM: {0}, SpO2: {1}".format(self.bpm, spo2))
                            self.print_spo2 = spo2
                            self.print_bpm = self.bpm

            time.sleep(self.LOOP_TIME)

        sensor.shutdown()

    def start_sensor(self):
        self._thread = threading.Thread(target=self.run_sensor)
        self._thread.stopped = False
        self._thread.start()

    def stop_sensor(self, timeout=2.0):
        self._thread.stopped = True
        self.bpm = 0
        self._thread.join(timeout)
```

Step 3: Configuration of Library for MLX90614.

Download the library from the link: https://pypi.org/project/PyMLX90614/#files. Before using the library, it must be extracted to the same folder which was created earlier. There are no changes needed in this library.

Step 4: Configuration of ThingSpeak Channel

To create ThingSpeak channel, one has to first sign up on ThingSpeak (https://thingspeak.com/). In case one has an account on ThingSpeak just sign in using your id and password. For signup, fill your details, then verify with the received email and proceed. After this, click on 'New Channel' button which will subsequently ask for the 'Name and Description' of the data you want to upload on this channel as shown in Fig. 5.6.

In this application Pulse Rate, Oxygen Saturation, and Body Temperature are sent to ThingSpeak cloud. Hence, the authors have named the channel as "Pulse, Oxygen Saturation, and Body Temperature".

More than one field of data can be activated by checking the box next to the 'Field' option as shown in Fig. 5.6. The authors have created three fields, namely, BPM, SpO2, and Temp. After this, click on 'save channel' button to save the details.

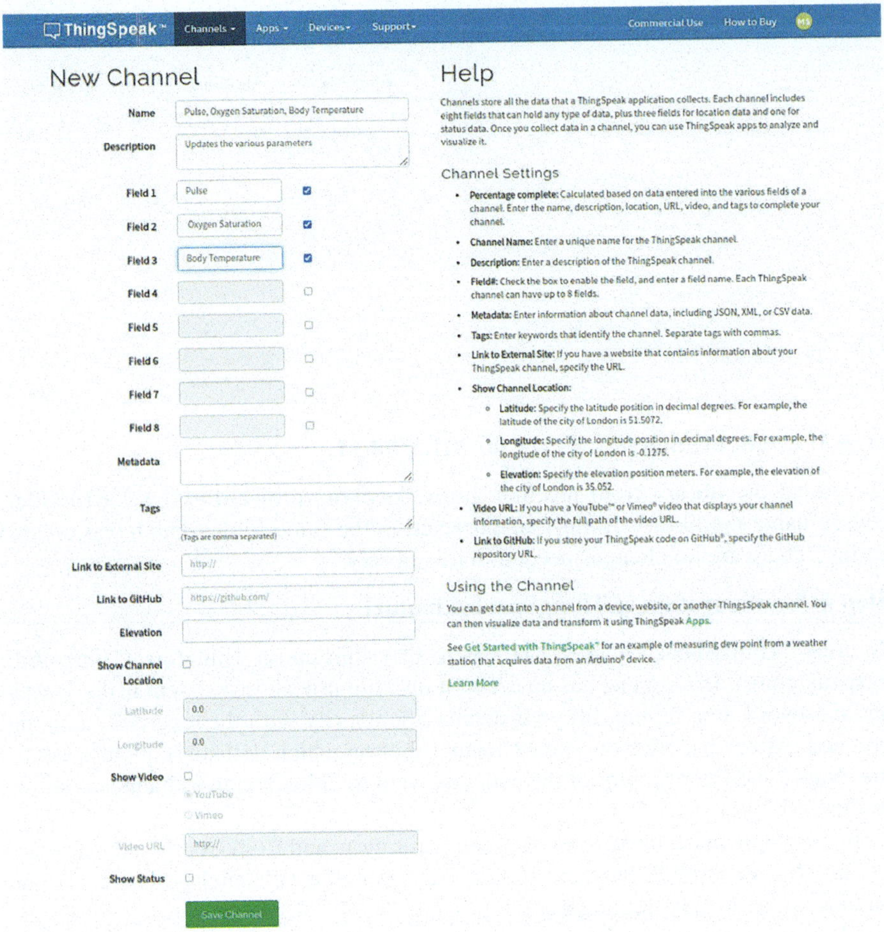

Fig. 5.6 Name and description of channel for Patient Monitoring system

Obtain the API Key

To send data to ThingSpeak, a unique API key is required, which is used in the main code to upload body parameters to ThingSpeak server. Navigate to 'API Keys' header under the newly created above channel to get a unique API key as shown in Fig. 5.7.

Now copy '*Write API Key*' to use it in the main code. To set up the Python environment for sending data to ThingSpeak, one must install libraries using the below commands.

```
sudo apt-get install httplib
sudo apt-get install urllib
```

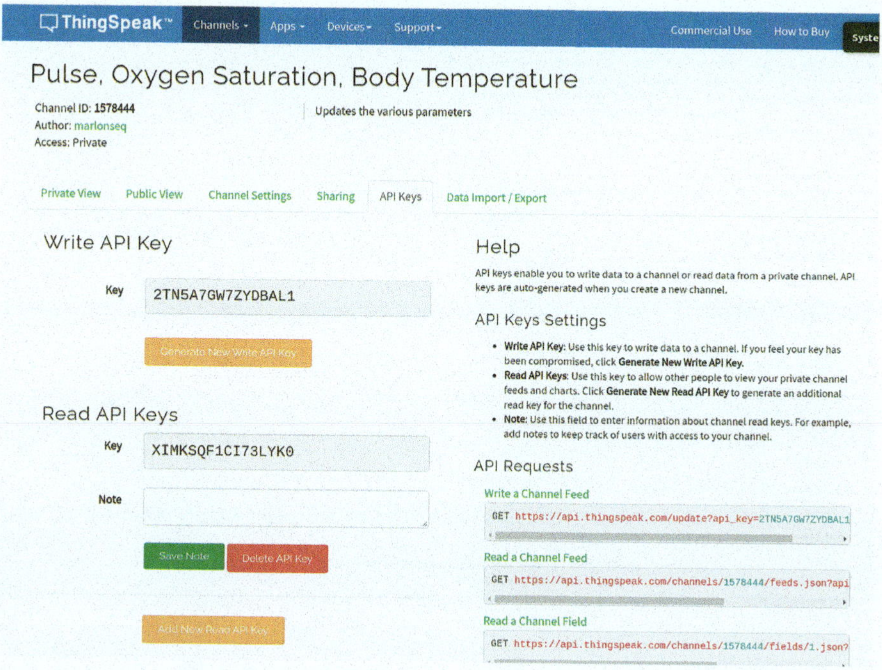

Fig. 5.7 API Key for Patient Monitoring system

After completing these steps the channel is ready to send the body parameters. Fig. 5.8 gives the flow chart of Patient Monitoring System with Cloud.

Step 4:

Create a file 'main.py' in the project folder with the below code.

```
# Imports required for MAX30102 sensor
from heartrate_monitor import HeartRateMonitor
import time
import argparse

# Imports required for MLX90614 sensor
import smbus
from mlx90614 import MLX90614

# Imports required for Thingspeak cloud update
import os
import http.client
import urllib

key = "2TN5A7GW7ZYDBAL1"  # Put your Thingspeakcloud API Key here
```

```
# MAX30102 sensor update section

parser = argparse.ArgumentParser(description="Read and print data from MAX30102")
parser.add_argument("-r", "--raw", action="store_true",
                    help="print raw data instead of calculation result")
parser.add_argument("-t", "--time", type=int, default=30,
                    help="duration in seconds to read from sensor, default 30")
args = parser.parse_args()


print('sensor starting...')
hrm = HeartRateMonitor(print_raw=args.raw, print_result=(not args.raw))
hrm.start_sensor()
try:
    time.sleep(args.time)
except KeyboardInterrupt:
    print('keyboard interrupt detected, exiting...')

hrm.stop_sensor()
print('sensor stoped!')
print(hrm.print_bpm,hrm.print_spo2)

# MLX90614 sensor update section

bus1 = smbus.SMBus(1)
sensor = MLX90614(bus1, address=0x5a)
print ("bodyTemperature :", sensor.get_object_1())
body_temp = sensor.get_object_1()
bus1.close()

# Thingspeak cloud upload section

params = urllib.parse.urlencode({'field1': hrm.print_bpm, 'field2': hrm.print_spo2, 'field3': body_temp, 'key':key })


headers = {"Content-typZZe": "application/x-www-form-urlencoded","Accept": "text/plain"}
conn = http.client.HTTPConnection("api.thingspeak.com:80")
conn.request("POST", "/update", params, headers)
response = conn.getresponse()

print(response.status,response.reason)
data = response.read()
conn.close()
```

In the Heart Rate and SpO2 section of the code, authors instantiate an object of the class '*HeartRateMonitor*' found in '*heartrate_monitor.py*'. The thread is initiated and terminated by calling '*start_sensor*' and '*stop_sensor*', respectively. When the thread is active one can access '*bpm*' to get beats per minute (BPM). The user has to wait for a few seconds to get a reliable BPM value. Also, make sure that there is minimum movement as the sensor is very sensitive to it. The program is designed in such a way that a keyboard interrupt (CNTRL + C) can stop the sensor and transit to the next section of the code. The last reading for Pulse rate and Oxygen saturation is written into '*hrm.print_bpm*' and '*hrm.print_spo2*' variables, respectively, which are then sent to ThingSpeak Cloud.

Step 5: To run the above '*main.py*' use the below command.

```
python3 main.py
```

The output after execution is shown on the terminal as in Fig. 5.9 and also the data visualization on the ThingSpeak server is shown in Fig. 5.10.

5.6.2 *Home security with FoG*

Certainly, the advent of the IoT and FoG computing paradigm has made it available for user to efficiently approach the problem of Home security. The authors have leveraged

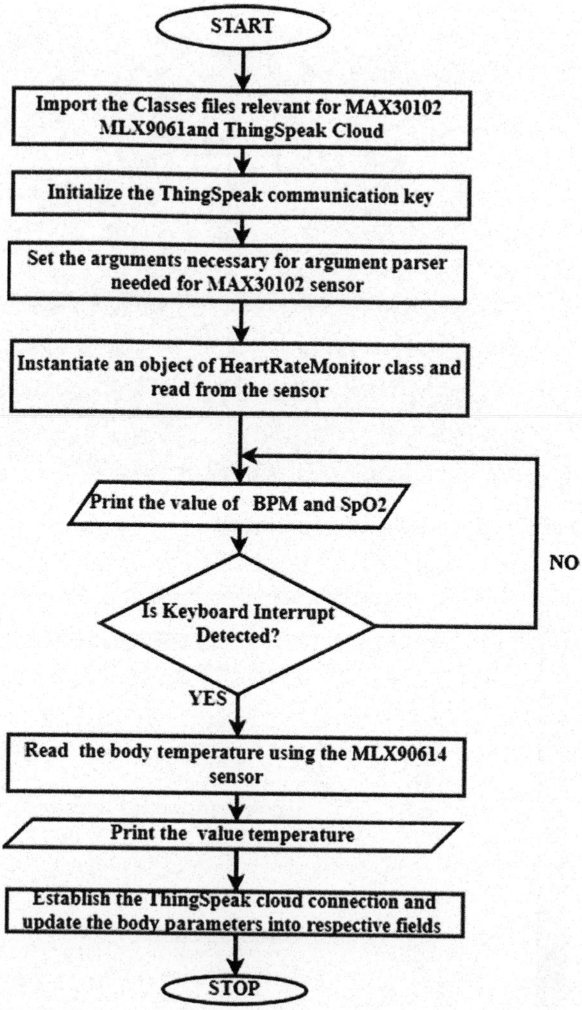

Fig. 5.8 Flowchart for Patient Monitoring system with cloud

various security hardware such as PIR sensors, Surveillance cameras, Smart locks, etc. to safeguard property and human life from potential life threatening situations. In this section, authors have implemented Home Surveillance, Home Safety Lock, and Fire alert system. The programming of the IoT nodes is done by Thonny IDE using MicroPython programming as detailed in below steps.

Step 1: Download the Thonny IDE for windows from the link: https://thonny.org

Step 2: Download ESP 32 Firmware from the link https://micropython.org/download/esp32/

Step 3: Install appropriate MicroPython Firmware on ESP32:

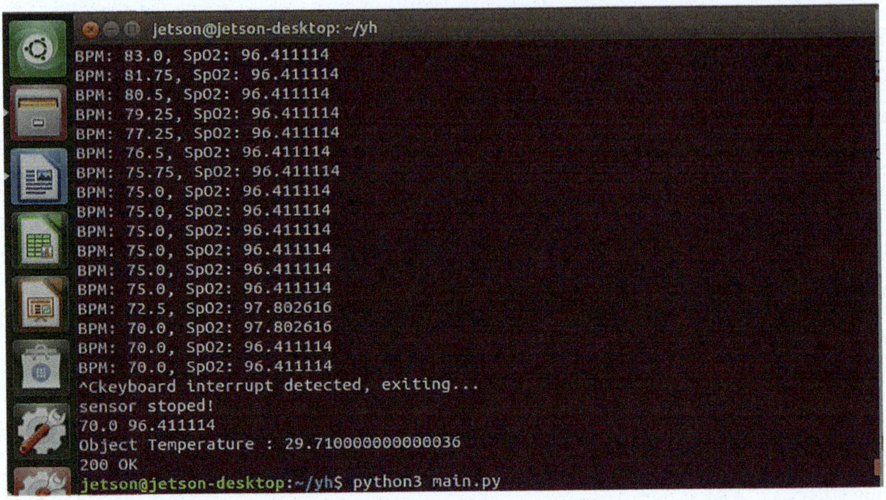

Fig. 5.9 Health parameters on terminal

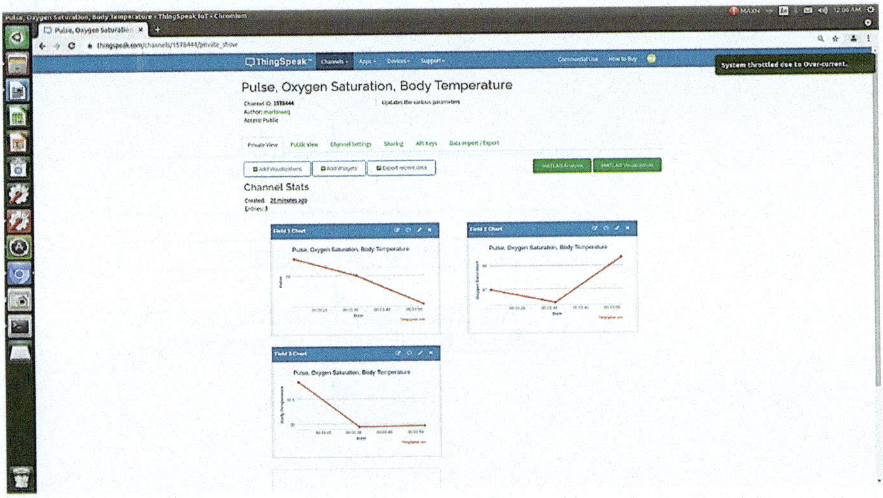

Fig. 5.10 Health parameters on ThingSpeak cloud

Open Thonny IDE → Tools → options → interpreter → select MicroPython (ESP32) → select COM port (Make sure the virtual com port drivers are installed) → click on “'install and update firmware' → select port and firmware path (downloaded MicroPython Firmware path) → then click on 'install'.

Step 4: MicroPython interpreter selection:

Goto Tools → options → click on interpreter tab and select drop down for Micropython (ESP 32). Also select the proper COM port for communication.

5.6.2.1 Home Surveillance

Home Surveillance system consists of IoT device ESP32-CAM having PIR sensor interfaced to it. IoT device communicates with Jetson Nano SBC which is configured as FoG node as shown in Fig. 5.11. Intruder activity is sensed by a PIR sensor and an image of the activity is captured by a camera interfaced with IoT device. The captured image is sent over Wi-Fi using MQTT protocol to the Jetson Nano SBC. The Jetson Nano SBC(FoG node) stores the received image. The Jetson upon reception of the image will intimate the house owner with a prompt WhatsApp message to his mobile phone using the Twilio cloud python package. The FoG node will also upload the same image to the Google Drive cloud so that the house owner can monitor the house in real time. The detailed conceptual and pin diagram are shown in Fig. 5.11. Table 5.2 depicts the pin connection of PIR sensor to ESP32 CAM.

IoT Device (ESP32-CAM)

The ESP32-CAM is a fully-featured microcontroller that has an integrated video camera and microSD card socket. It is easy to use, and perfect for various IoT

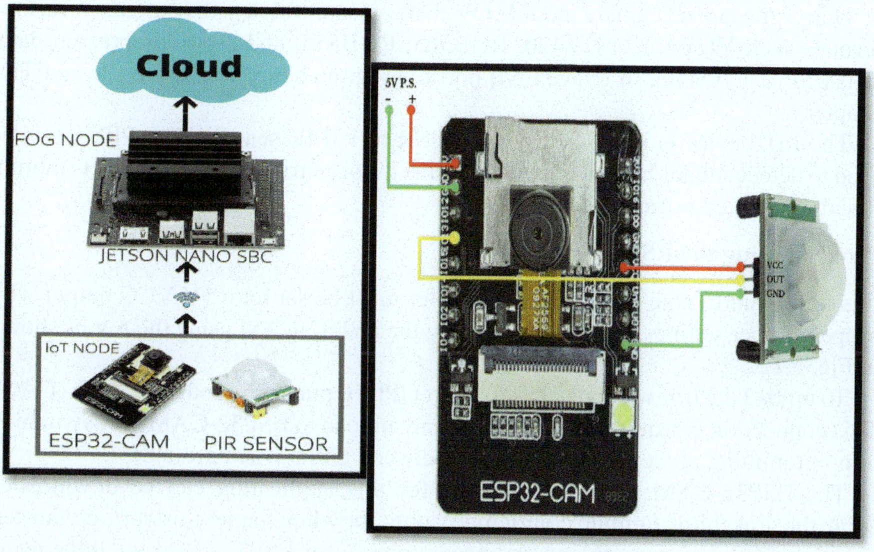

Fig. 5.11 Conceptual and pin diagram of home surveillance system

Table 5.2 Pin connection for Home Surveillance

ESP32-CAM	PIR sensor
VCC	VCC
GND	GND
IO13	OUT

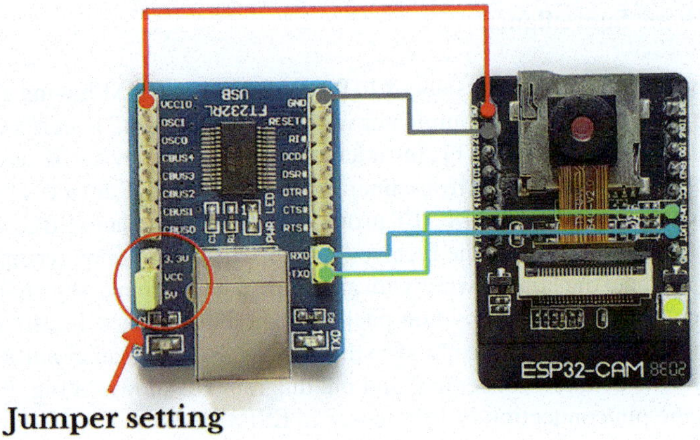

Fig. 5.12 ESP-32 CAM programming through FTDI 232

applications requiring a camera with advanced functions like image tracking and recognition.

Here,2 megapixel camera module OV2640 is used, which gives outputs in various formats such YUV422, YUV420, RGB565, RGB555, and 8-bit compressed data. This ESP32-CAM doesn't have USB port to program hence one has to use an FTDI adapter.

The IoT device is interfaced with HC-SR 501 (PIR sensor). The PIR sensor is used to detect animal/human movement. This pyroelectric PIR sensor detects motion based on emitted infrared radiation.

Programming the ESP32-CAM:

It is important to note that the FTDI adapter must be set for a 5V VCC output with proper jumper settings, as we are powering the ESP32-CAM using the 5 V as shown in Fig. 5.12.

To upload the firmware, one has to short GPIO 0 pin to Ground on ESP32 - CAM. This connection is required only for firmware upload to ESP32-CAM. For switching to programming mode remove the short between GPIO 0 and ground.

The ESP32-CAM firmware required for this application can be downloaded from the link https://github.com/lemariva/micropython-camera-driver/tree/master/firmware. The firmware upload and programming for ESP32-CAM are done using MicroPython in the Thonny IDE.

In order to configure ESP32-CAM for MQTT protocol, one requires MicroPython MQTT client(umqtt) and camera package which can be downloaded using the steps outlined in Sect. 3.2 of Chapter 3. Now the ESP32-CAM is ready to capture the image and implement the MQTT protocol. Flowchart for configuring ESP32-CAM module for Home security is shown in Fig. 5.13.

Create a 'main.py' file in Thonny IDE and enter the below code in it. Then click the 'run' button to execute the code. Make sure before running the code, the Wi-Fi

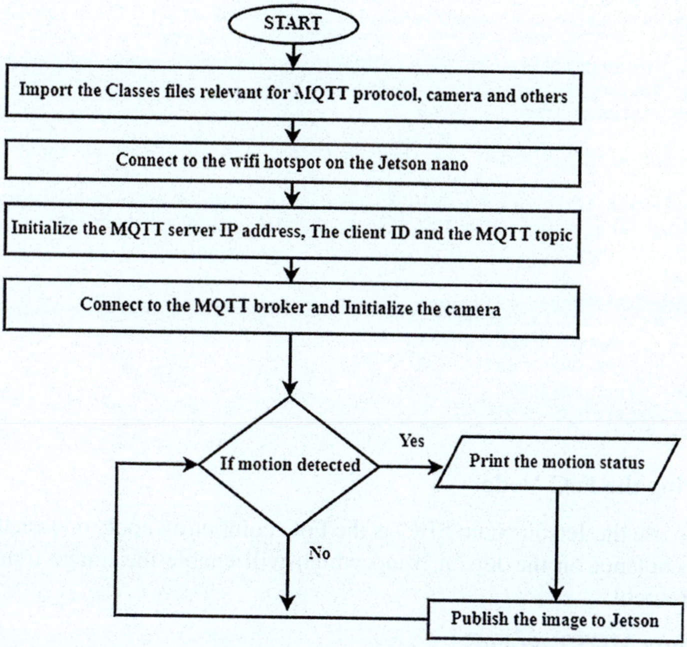

Fig. 5.13 Flowchart for configuring ESP32s-CAM module for Home security

hotspot created on FoG node is active. Here ESP32-CAM is the client which will publish the captured image to MQTT topic named as 'Image'.

```
from machine import Pin          #importing classes
from umqtt.simple import MQTTClient
import time
import os
import camera

# make sure wifi hotspot is enabled on jetson nano SBC(FoG Node)
import network
station = network.WLAN(network.STA_IF)
station.active(True)
station.connect("jetson-desktop", "11111111")

SERVER = '10.42.0.1'  # MQTT Server Address (IP address of jetson hotspot)
CLIENT_ID = 'esp32-camera'
TOPIC = b'Image'

# Connect to MQTT broker
c = MQTTClient(CLIENT_ID, SERVER)
c.connect()

camera.init(0, format=camera.JPEG)
```

```
Motion_status = False   #Global variable to hold the state of motion sensor

def handle_interrupt(Pin):             #defining interrupt handling function
    global Motion_status
    Motion_status = True

PIR_Interrupt=Pin(13,Pin.IN)   # setting GPIO13 PIR_Interrupt as input

#Attach external interrupt to GPIO13 and rising edge as an external event source
PIR_Interrupt.irq(trigger=Pin.IRQ_RISING, handler=handle_interrupt)

while True:
    if Motion_status:
       print('Motion is detected!')
       buf = camera.capture()
       c.publish(TOPIC, buf)
       time.sleep_ms(100)
       Motion_status = False
```

Configuring the FoG Node

In order to use the Jetson Nano SBC as the FoG computing node, one need to enable the Wi-Fi hotspot on the Jetson Nano which will enable the image transfer using MQTT protocol.

Establishing MQTT on FoG:

MQTT is a lightweight publish-subscribe network protocol that transports messages between devices. The MQTT protocol defines two types of network entities: a broker and clients. An MQTT broker is a server responsible to receive and route all messages from the clients to the appropriate destination. An MQTT client is any device that connects to an MQTT broker over a network.

For MQTT on the FoG Node, Mosquitto library package is required to be installed. This can be done by running the below commands on the Jetson terminal.

```
sudo apt-get install mosquittomosquitto-clients
```

One also requires the '*paho-mqtt*' library , which can be installed using the below command.

```
sudo apt-get install python3-pip
sudo pip3 install paho-mqtt
```

The Jetson FoG node is now ready for MQTT protocol communication.

Enabling Wi-Fi Hotspot on FoG node:

To enable the hotspot on the FoG node, select the settings menu on Jetson user interface. Under setting menu → Select 'Wi-Fi' → on the right-hand side, hover over the three horizontal lines (≡) → Turn on Wi-Fi hotspot.

It is not possible to access the Internet over wireless when hotspot is active. Hence a wired connectivity is provided for the Jetson Nano SBC to connect to the Internet for uploading the image to Google Drive and sending a message using Twilio cloud.

Configuring the FoG Node to Communicate with Cloud

In this application, authors have used two cloud platforms, namely, Twilio and Google Drive cloud. The Twilio communication platform is used to send the WhatsApp message indicating that there is an intruder activity. At the same time, the surveillance image is also uploaded to the Google Drive cloud. The configuration details of both these platforms are given below.

Twilio Configuration:

Next, one needs to prepare the Jetson Nano SBC (FoG node) to send the WhatsApp message using Twilio cloud communication platform. One has to create the Twilio account by visiting the website (https://www.twilio.com/)

Click on the signup option, and fill the respective user details. Once the account is created the verification of registered email and phone number is done.

Once the verification steps are done, one needs to specify the Twilio features required. Select the features using dropdown and radio buttons as shown in Fig. 5.14. At last click on button 'Get started with Twilio' which will take user to the console, wherein user needs to agree to activate the sandbox and click confirm.

Under the 'Manage account' navigate to 'general settings'. Next copy the Account SID and the Auth token which are highlighted in Fig. 5.15. These are needed in python code to send a WhatsApp message using the Twilio cloud.

Next step is to send a WhatsApp message with content "'join every-plain' to the phone number "+14155238886" (same number must be entered) from the registered phone number which will be used to receive the notification.

Setting up Jetson Nano SBC (FoG node) for Twilio

Before using the python code to receive and send WhatsApp messages we need to install the Twilio package on Jetson Nano SBC using the following command:

```
sudo pip install twilio
```

Now the system is ready to send and receive notifications on WhatsApp via the Twilio cloud platform.

Google Drive Configuration

In this implementation, the surveillance image is stored on to Google Drive. To do this, user need to install the Pydrive Python module using the following command.

```
pip install pydrive
```

To upload the image to Google Drive using python, one need an active google cloud account. User need to get the authentication files for Google Service API, so that the Python code can access Google Drive. To do that, one need to follow the below steps.

Step 1: Create a new project in Google Developer Console by clicking 'CREATE PROJECT' with the proper name of the project. The link for the

Fig. 5.14 Twilio feature selection

same is https://console.cloud.google.com/cloud-resource-manager?organizationId=0&authuser=1&supportedpurview=project.

Step 2: Next step is to Enable APIs and their Services by clicking the 'ENABLE APIS AND SERVICES'. This will bring you to the API library. Then Search 'Google Drive' → Click the 'Google Drive API' icon → Then click 'ENABLE', which will enable Google Drive API service.

Step 3: Creation of credentials: click on 'credentials' and select the options shown in Fig. 5.16 and click 'done'. Next hover over to 'create credentials' and select 'OAuth Client ID' as shown in Fig. 5.17.

Step 4: Next click on the button 'configure consent screen' → Navigate to OAuth consent screen → select 'External' and then click 'Create'.

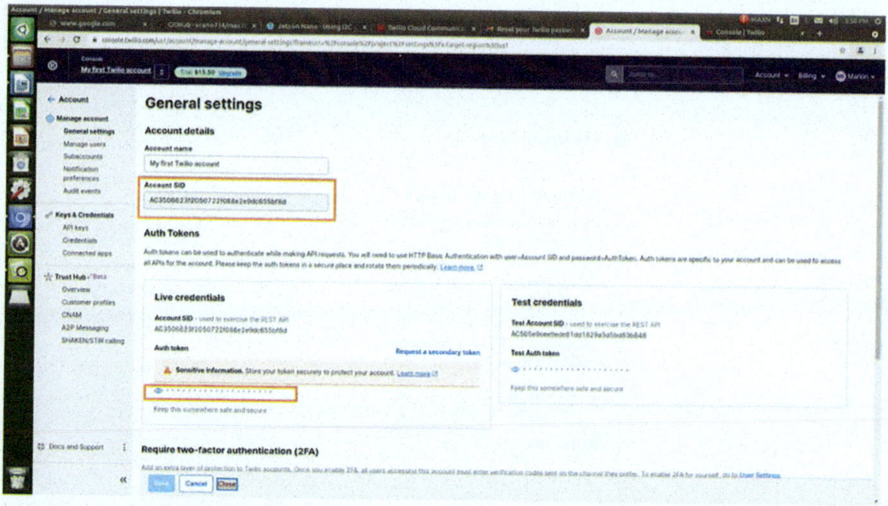

Fig. 5.15 Account SID and Auth token

On the next screen enter the 'app name' and the 'user support email id' as well as the developer contact information (developer email id) and then click 'save' and 'continue'. On the following screen, click on 'save and continue' as no changes are required. Next add a test user's email id and click 'add' → then click 'save and continue'. The next screen displays the summary of the OAuth consent.

Step 5: Next under Credential on left panel click 'Create Credentials' → select OAuth Client ID → Provide the settings as shown in Fig. 5.18 and click 'create'. Now the OAuth Client ID is created.

Step 6: After creating the OAuth client ID → click on 'download' button to get .JSON file. This downloaded JSON file is required for Python code to access Google Drive. After downloading the .JSON file, rename it to *'client_secrets.json'* and make sure that it is placed in the same folder where the 'main.py' is present.

Step 7: The Python code for home surveillance running on Jetson Nano SBC (FoG node) requires the user to enter the google account and password. After authentication is completed browser displays *'The authentication flow has completed'* and initiates the uploading of the image onto Google Drive.

The surveillance image sent by ESP32-CAM is received by FoG node will be saved in the folder named as 'photos' in the project folder of the FoG node. The same image is also uploaded to Google Drive in its root directory for remote monitoring. The flow chart for configuring Jetson Nano for Home security is given in Fig. 5.19.

Step 8: Create a 'main.py' to enter the below Python code on the Jetson Nano SBC and run it using the below command.

```
python3 main.py
```

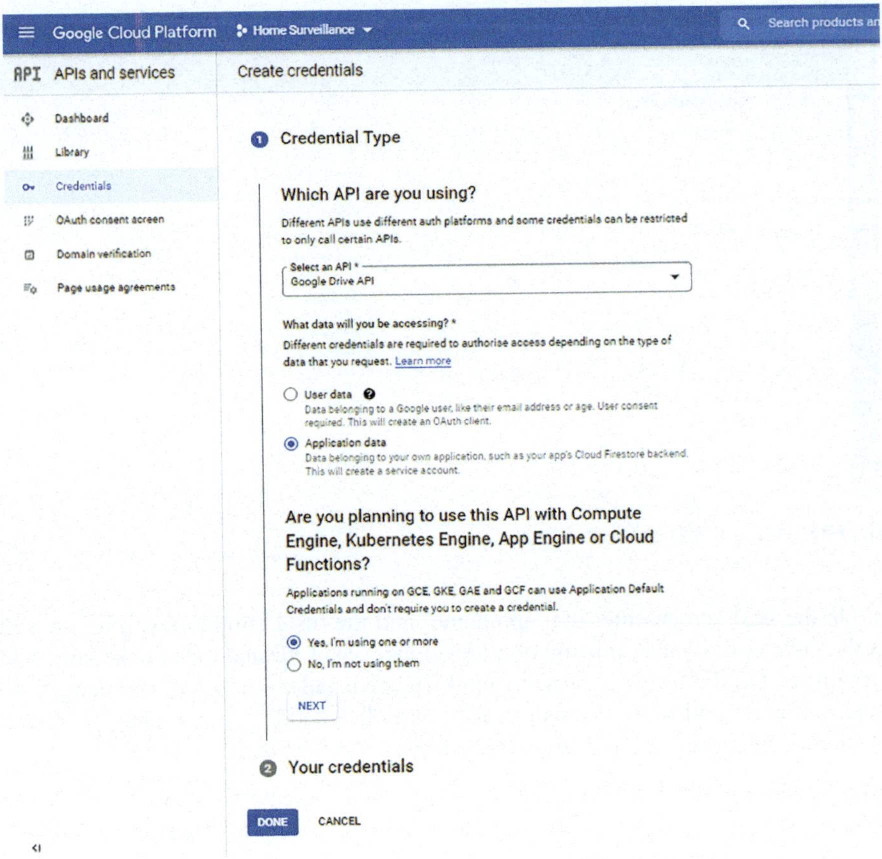

Fig. 5.16 Creation of credentials

After successfully running the code the Twilio message notification is sent to the registered mobile number of the house owner as shown in Fig. 5.20.

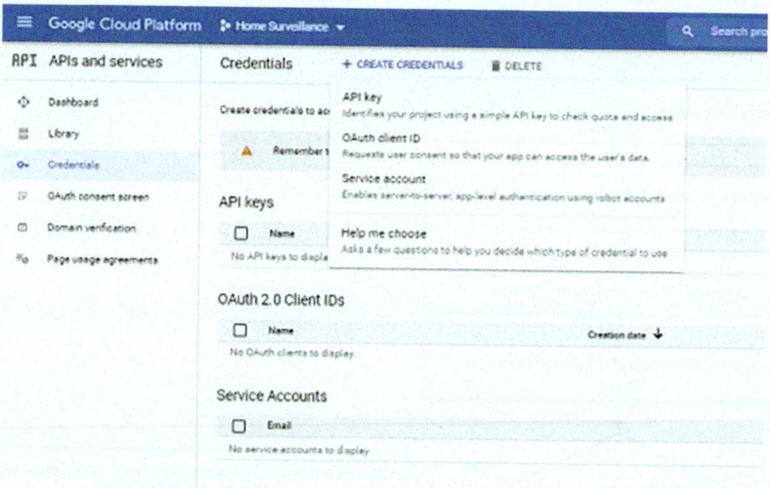

Fig. 5.17 OAuth client ID

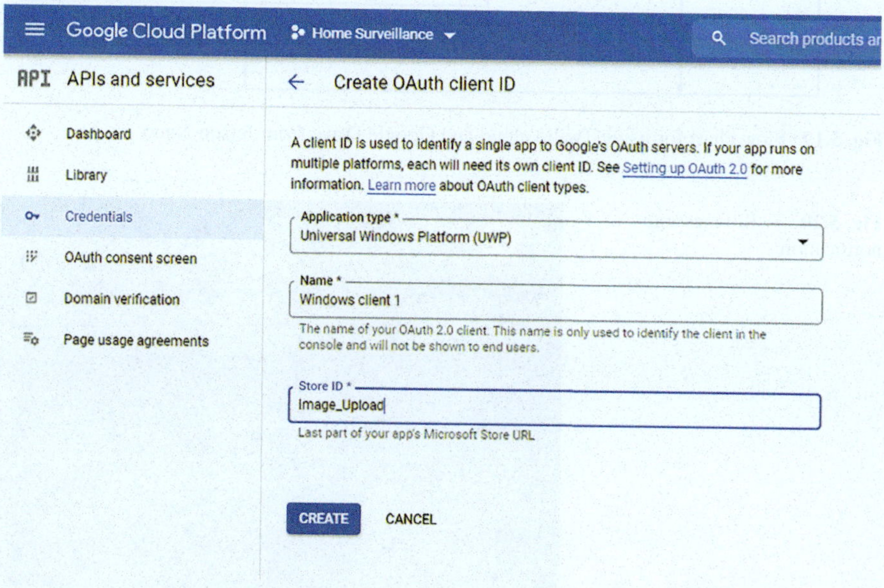

Fig. 5.18 OAuth client ID creation

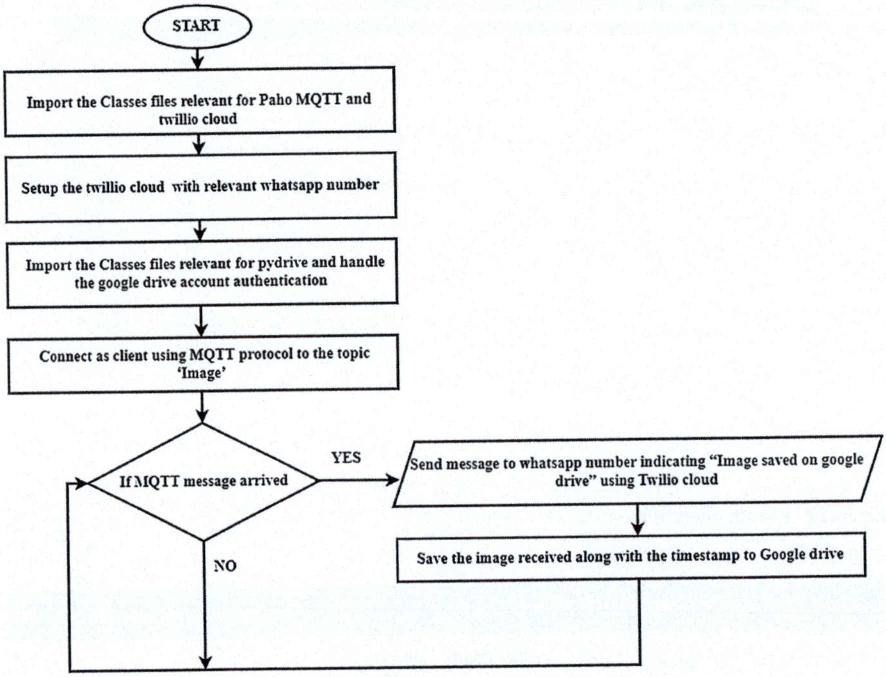

Fig. 5.19 Flow chart for using Twilio cloud and Google Drive from Jetson Nano

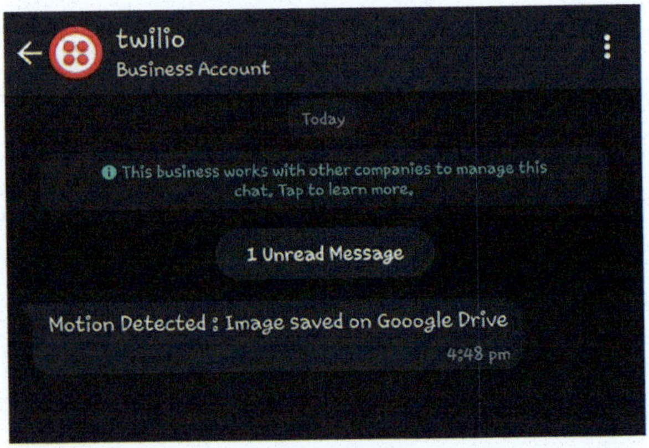

Fig. 5.20 Twilio message notification

```
import time
import paho.mqtt.client as mqtt

#twilio imports
import os
from twilio.rest import Client

# setuptwilio

twilio_client = Client('AC230101cb426d251b82479e33a992c149', '1d55483a6ab87ed20eac8531d70650c4');
# this is the Twilio sandbox testing number
from_whatsapp_number='whatsapp:+14155238886'

# replace this number with your personal WhatsApp Messaging number
to_whatsapp_number='whatsapp:+91705750xxxx'

# setuptwilio ends

# Google drive setup
from pydrive.drive import GoogleDrive
from pydrive.auth import GoogleAuth

# Google drive authentication
gauth = GoogleAuth()


# Creates local webserver to handle authentication
gauth.LocalWebserverAuth()
drive = GoogleDrive(gauth)

# Replace with the path, where the image is present
path = r"/home/jetson/MQTT-M5Camera/photos"

# Googel drive setup ends here

def on_connect(client, userdata, flags, rc):
    print("Connected with result code "+str(rc))
    client.subscribe('Image')

def on_message(client, userdata, msg):
    # generate filename
    timestamp = time.gmtime()
    time_str = '%4d%02d%02d%02d%02d%02d' %(timestamp[0], timestamp[1], timestamp[2], timestamp[4], timestamp[5], timestamp[6])
    # Create a file with write byte permission
    f = open('photos/'+time_str+'.jpg', "wb")
    f.write(msg.payload)
    f.close()
    print("Image received and saved!")

    # twilio message
    message = twilio_client.messages.create(body='Motion Detected : Image saved on GooogleDrive',from_=from_whatsapp_number,to=to_whatsapp_number)
    print(message.sid)

    # Google drive image upload

    t = drive.CreateFile({'title': time_str+'.jpg'})
    t.SetContentFile(os.path.join(path, time_str+'.jpg'))
    t.Upload()

client = mqtt.Client()
client.on_connect = on_connect
client.on_message = on_message
client.connect('localhost', 1883, 60)

client.loop_forever()
```

5.6.2.2 Home Safety Lock

Safety is the most important requirement for any home. More than ever before, a need is felt for a home safety lock to safeguard one's house and the belonging therein. With the advent of IoT, one can give the connectivity to the door lock.

Here, authors have designed a smart home safety lock that communicates with the owner. It also informs the owner of an authentic entry into the house, or if an intruder is trying to enter the house. The solenoid lock is placed with IoT node configured on

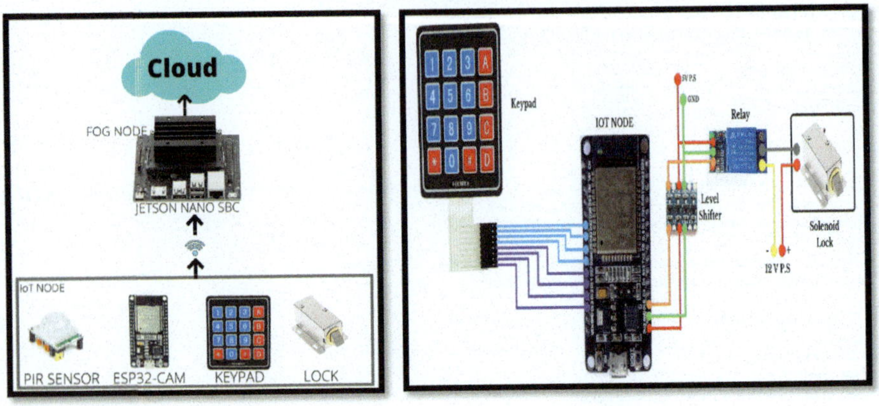

Fig. 5.21 Conceptual and pin diagram for home safety lock

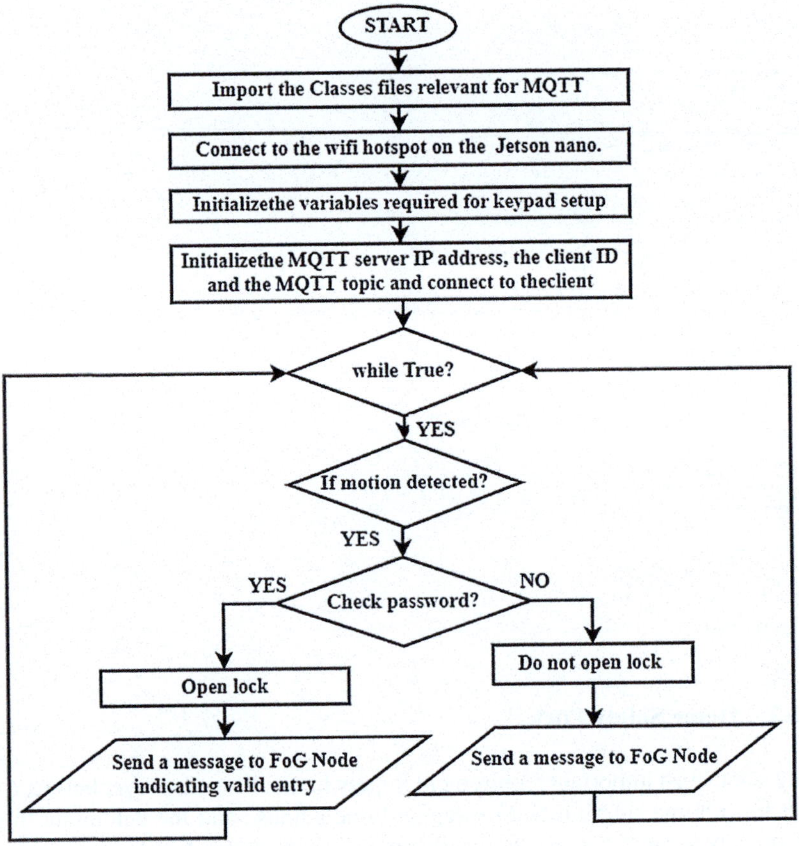

Fig. 5.22 Flow chart for home safety lock on ESP32 Wi-Fi module

Table 5.3 Pin connection for Home Safety lock

ESP3 Wi-Fi module	Keypad	Level shifter input
VCC	–	–
GND	–	–
3V3	–	LV
GND	–	GND
D15	–	LV1
D13	ROW 1	–
D12	ROW 2	–
D14	ROW 3	–
D27	ROW 4	–
D26	COL 1	–
D23	COL 2	–
D33	COL 3	–
D32	COL 4	–

Level shifter output	Relay module
HO	5 V(power supply)
HO1	IN
GND	GND

Relay module	Solenoid lock	12 V Power supply
–	RED PIN	+ PIN(RED)
NO	–	- PIN(BLACK)
COM	GND PIN	–

ESP32 Wi-Fi module. When PIR sensor detects motion, the IoT node prompts the user to enter the password. If the password is valid, only then the lock will open and the user is allowed to enter the house. The status of valid/invalid entry is informed to the Jetson Nano SBC configured as FoG node using MQTT protocol. The FoG node in turn sends an alert message to the house owner using Twilio cloud platform on WhatsApp. The conceptual diagram and pin diagram is shown in Fig. 5.21. Table 5.3 depicts the pin connection for the home safety lock.

To Configuring MQTT on ESP 32 Wi-Fi Module:

To configure, follow the steps already given in Sect. 5.6.2.1. This includes connecting to the hotspot hosted by Jetson Nano SBC.

Here, ESP32 is the client and publishes the sensor data to a topic called 'password_verify'. FoG node, i.e. the Jetson Nano SBC will be the broker and client. A Python MQTT client running on the FoG node will subscribe to the 'password_verify' topic and collect the results. After setting up the MQTT protocol on the IoT node, it is ready for communication. The flow chart for configuring ESP32 for Home Safety Lock is given in Fig. 5.22.

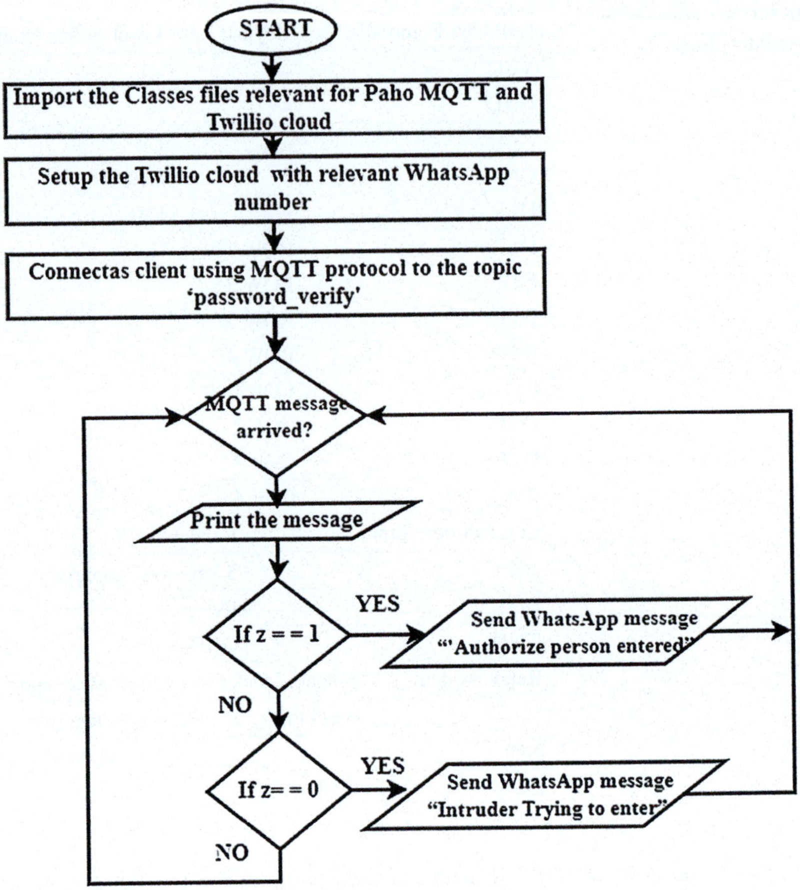

Fig. 5.23 Flow chart for configuring Home Safety Lock using Jetson Nano

Create a 'main.py' file in Thonny IDE and enter Python code. Then click the 'run' button to execute the code.

```
from machine import Pin        #importing classes
from time import sleep         #Import sleep from time class
from umqtt.simple import MQTTClient
import time
import machine

# make sure wifi hotspot is enabled on jetson nano SBC(FoG Node)
import network
station = network.WLAN(network.STA_IF)
station.active(True)
station.connect("jetson-desktop", "11111111")

KEY_UP   = const(0)
KEY_DOWN = const(1)

keys = [['1', '2', '3', 'A'], ['4', '5', '6', 'B'], ['7', '8', '9', 'C'], ['*', '0', '#', 'D']]

# Pin names
rows = [13,12,14,27]
cols = [26,25,33,32]

# set pins for rows as outputs
row_pins = [Pin(pin_name, mode=Pin.OUT) for pin_name in rows]

# set pins for cols as inputs
col_pins = [Pin(pin_name, mode=Pin.IN, pull=Pin.PULL_DOWN) for pin_name in cols]


def init():
    for row in range(0,4):
        for col in range(0,4):
            row_pins[row].value(0)

def scan(row, col):
    """ scan the keypad """
    # set the current column to high
    row_pins[row].value(1)
    key = None

    # check for keypressed events
    if col_pins[col].value() == KEY_DOWN:
        key = KEY_DOWN

    if col_pins[col].value() == KEY_UP:
        key = KEY_UP

    row_pins[row].value(0)
    # return the key state
    return key


SERVER = '10.42.0.1'  # MQTT Server Address (Change to the IP address of your Pi)
CLIENT_ID = 'ESP32'
```

```
TOPIC = b'password_verify'

client = MQTTClient(CLIENT_ID, SERVER)
client.connect()    # Connect to MQTT broker


Motion_status = False  #Global variable to hold the state of motion sensor

def handle_interrupt(Pin):           #defining interrupt handling function
    global Motion_status
    Motion_status = True

lock=Pin(15,Pin.OUT)   #setting GPIO14 led as output
PIR_Interrupt=Pin(34,Pin.IN)    # setting GPIO13 PIR_Interrupt as input

#Attach external interrupt to GPIO13 and rising edge as an external event source
PIR_Interrupt.irq(trigger=Pin.IRQ_RISING, handler=handle_interrupt)

password = ['1','0','0','0']
input_pass = ['0','0','0','0']


def password_verify():           #defining interrupt handling function
        print('Please enter the password')
        break_loop = 0
        i = 0
            while i<4:
                while break_loop == 0:
                    for row in range(4):
                        for col in range(4):
                            key = scan(row, col)
                            if key == KEY_DOWN:
                                print("Key Pressed", keys[row][col])
                                time.sleep(0.5)
                                last_key_press = keys[row][col]
                                break_loop = 1
                break_loop = 0
                input_pass[i]=last_key_press
                i=i+1



# set all the columns to low
init()
lock.value(1)# initially locked
while True:
    if Motion_status:
        print('Motion is detected!')
        password_verify()

        print(input_pass)

        if input_pass == password:
            print('yes')
            lock.value(0)
```

```
            t = 1.0    # 1.0 indicates authentication done,lock opened
            if isinstance(t, float):  # Confirm sensor results are numeric
                msg = (b'{0:3.1f}'.format(t))
                client.publish(TOPIC, msg)  # Publish data to MQTT
                print(msg)
            else:
                print('Invalid sensor readings.')

            sleep(10)
            lock.value(1)
            Motion_status = False
        else:
            print('No')
            lock.value(1)   # dont open door
            t = 0.0         # 0.0 indicates authentication not done
            if isinstance(t, float):  # Confirm sensor results are numeric
                msg = (b'{0:3.1f}'.format(t))
                client.publish(TOPIC, msg)  # Publish sensor data to MQTT
                print(msg)
            else:
                print('Invalid sensor readings.')

            sleep(5)
            Motion_status = False
```

Upon a valid password entry, the IoT node will unlock the lock and at the same time send an MQTT message '1' to the Jetson Nano SBC (FoG Node) indicating the same. If the intruder tries to enter the invalid password, the IoT node will send an MQTT message as '0' to indicate that it is an invalid entry and will not open the lock.

To configure the Jestson Nano (FoG Node) for MQTT and Twilio cloud follow the steps given in Sect. 5.6.2.1. The flow chart for configuring Jetson Nano for Home Safety Lock is given in Fig. 5.23.

Create a 'main.py' and enter Python code and run using the below command.

```
python3 main.py
```

After successfully running the code, the Twilio message notification is sent to the registered WhatsApp mobile number of the house owner.

```
import paho.mqtt.client as mqtt
global z

#twilio imports
import os
from twilio.rest import Client

# setuptwilio

twilio_client = Client('AC230101cb426d251b82479e33a992c149','1d55483a6ab87ed20eac8531d70650c4')
# this is the Twilio sandbox testing number
from_whatsapp_number='whatsapp:+14155238886'

# replace this number with your personal WhatsApp Messaging number
to_whatsapp_number= 'whatsapp:+91705750xxxx'

# setup twilio ends
```

```
# Callback fires when conected to MQTT broker.
def on_connect(client, userdata, flags, rc):
    print('Connected with result code {0}'.format(rc))
    # Subscribe (or renew if reconnect).
    client.subscribe('password_verify')
# Callback fires when a published message is received.
def on_message(client, userdata, msg):
    print(msg)
    z = [float(x) for x in msg.payload.decode("utf-8").split(',')]
    if z[0] == 1.0:
        # twilio message
        message = twilio_client.messages.create(body='Authorized person
                                                entered',from_=from_whatsapp_number,
                                                to=to_whatsapp_number)

        print(message.sid)
    if z[0] == 0.0:
        message = twilio_client.messages.create(body='Intruder Trying to entered',
                                                from_=from_whatsapp_number,
                                                to=to_whatsapp_number)

        print(message.sid)


client = mqtt.Client()
client.on_connect = on_connect  # Specify on_connect callback
client.on_message = on_message  # Specify on_message callback
client.connect('localhost', 1883, 60)  # Connect to MQTT broker (also running on Jetson).

# Processes MQTT network traffic, callbacks and reconnections (Blocking)
client.loop_forever()
```

Once the code is executed on Jetson SBC (FoG node), it will receive a message from the IoT node using MQTT protocol. If the message received indicates an authentic entry, it will inform the owner of the same. If there is invalid entry, it indicates that the intruder is trying to enter.

5.6.2.3 Fire Alert System

Fire is one of the most dangerous threats that home and business owners need to take into account. In the unfortunate case of a fire, there is a risk of losing the majority of your belongings. A lack of safety precautions may result in material and financial losses, as well as death. As a result, having a fire alert system installed is a great way to keep premises safe. This method provides early alert of the hazard, allowing adequate time to evacuate the premises and contact authorities before the fire spreads out of control.

Here, the authors have designed a Fire Alert system to mitigate such situation. In this implementation, authors have used the ESP32-CAM module configured as IoT device. MQ-4 Sensor Module is interfaced to detect the smoke in the air. The MQ-4 can detect smoke concentrations anywhere from 200 to 10,000 ppm. It provides analog as well as digital output corresponding to the concentration of the gases in the air.

When the IoT node detects a fire, it will inform the Jetson Nano SBC (FoG node) using Wi-Fi over MQTT protocol. On reception of a message from the IoT node, the FoG node will inform the house owner by a WhatsApp message using the Twilio Cloud Platform, that there is a fire in the house for preventive measures. The

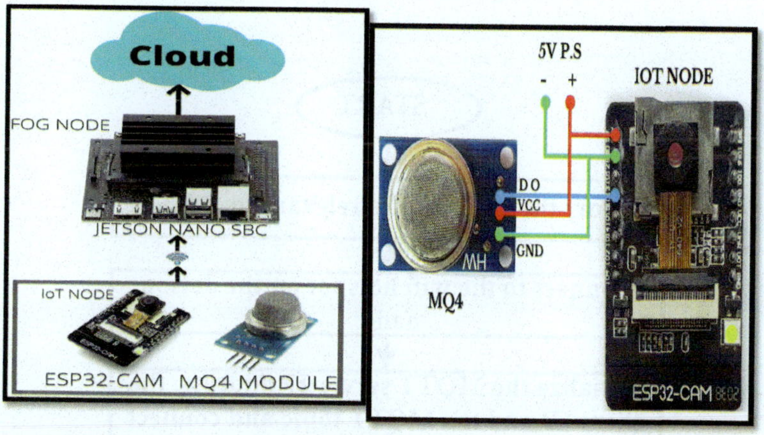

Fig. 5.24 Conceptual and pin diagram for Fire Alert system

Table 5.4 Pin connection for Fire Alert system

ESP32-CAM	MQ4 module
VCC	VCC
GND	GND
IO13	D O

conceptual diagram and pin diagram for Fire Alert system are shown in Fig. 5.24. Table 5.4 depicts the pin connection for the Fire Alert system.

To Configuring MQTT on ESP32-CAM

To configure ESP32-CAM module for MQTT follow the steps already given in Sect. 5.6.2.1. Here, ESP32 is the client and publishes the sensor data to a topic called 'smoke_alert'. FoG node, i.e. the Jetson Nano SBC will be the broker and client. A Python MQTT client running on the FoG node will subscribe to the 'smoke_alert' topic and collect the results. After setting up the MQTT protocol on the IoT node it is ready for communication. The flow chart for the above process is given in Fig. 5.25.

Create a 'main.py' file in Thonny IDE and enter the Python code . Then click the 'run' button to execute the code.

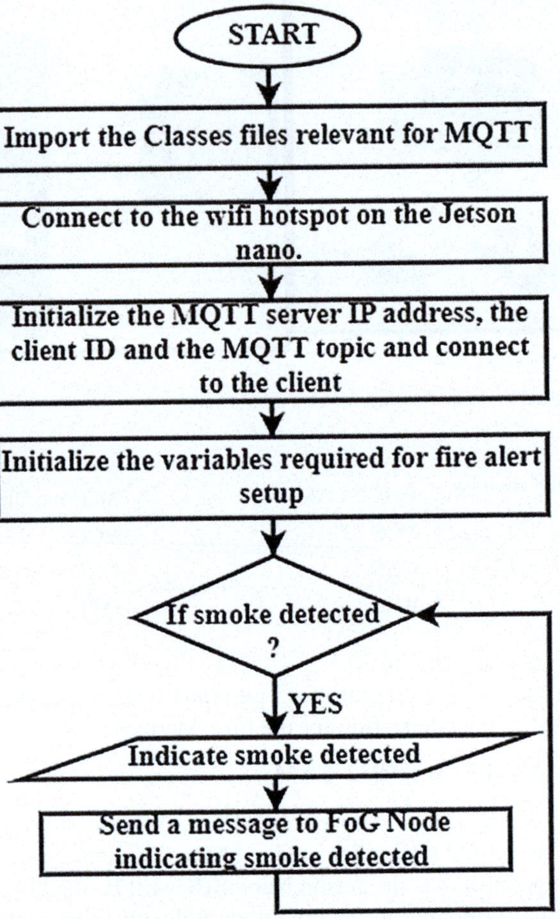

Fig. 5.25 Flow chart for configuring ESP 32 Wi-Fi module for Fire Alert system

```
import network
station = network.WLAN(network.STA_IF)
station.active(True)
station.connect("jetson-desktop", "11111111")


from machine import Pin        #importing classes
from time import sleep     #Import sleep from time class
from umqtt.simple import MQTTClient
import time
import machine


SERVER = '10.42.0.1'  # MQTT Server Address (Change to the IP address of your Pi)
CLIENT_ID = 'ESP32'
TOPIC = b'smoke_alert'

client = MQTTClient(CLIENT_ID, SERVER)
client.connect()   # Connect to MQTT broker


smoke_status = False  # variable to hold the state of motion sensor

MQ4=Pin(13,Pin.IN)   # setting GPIO13 as input


while True:
    if MQ4 value() == 0:
        smoke_status = True


    if smoke_status:
            print('smoke is detected!')

          t = 1.0          # 1.0 indicates authentication done
          if isinstance(t, float):  # Confirm sensor results are numeric
              msg = (b'{0:3.1f}'.format(t))
              client.publish(TOPIC, msg)  # Publish sensor data to MQTT topic
              print(msg)
          else:
              print('Invalid sensor readings.')
          sleep(5)
          smoke_status = False
```

To Configure the Jestson Nano (FoG Node) for MQTT and Twilio follow the steps given in Sect. 5.6.2.1. Now both the nodes are ready for communication. The flow chart for configuring Jetson Nano for the Fire Alert system is given in Fig. 5.26.

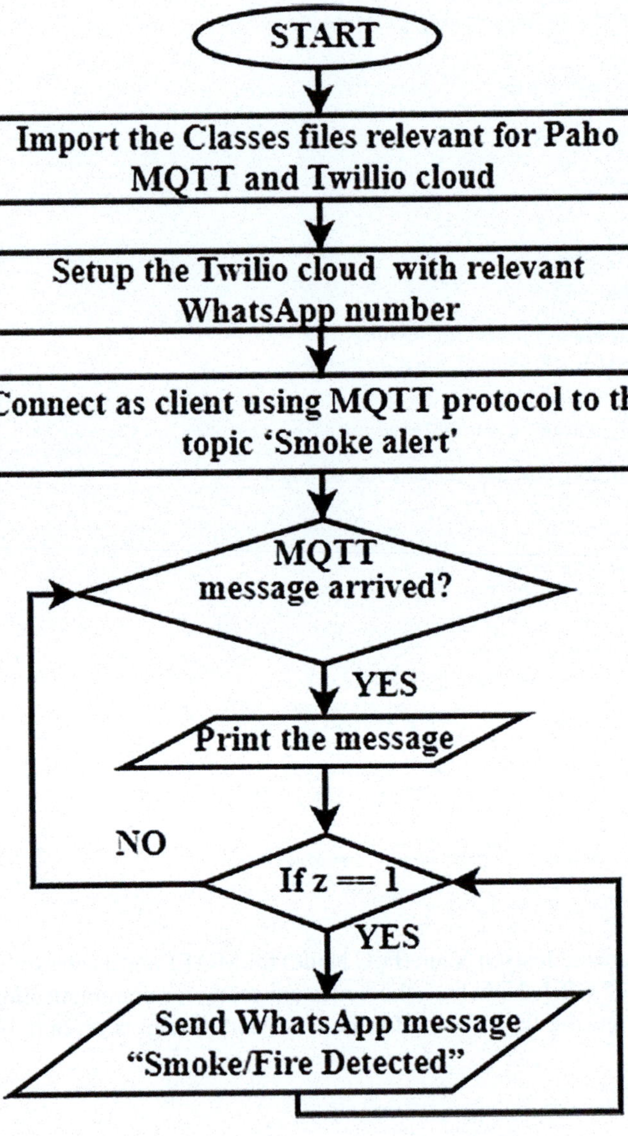

Fig. 5.26 Flow chart for configuring Jetson Nano for Fire Alert system

Next Create a 'main.py' with the below code on the Jetson Nano SBC and run it using the below command

python3 main.py

```
import paho.mqtt.client as mqtt
global z

#twilio imports
import os
from twilio.rest import Client

# setuptwilio
twilio_client =
Client('AC230101cb426d251b82479e33a992c149','1d55483a6ab87ed20eac8531d70650c4');
# this is the Twilio sandbox testing number
from_whatsapp_number='whatsapp:+14155238886'

# replace this number with your personal WhatsApp Messaging number
to_whatsapp_number='whatsapp:+917057504996'

# setuptwilio ends



# Callback fires when connected to MQTT broker.
def on_connect(client, userdata, flags, rc):
    print('Connected with result code {0}'.format(rc))
    # Subscribe (or renew if reconnect).
    client.subscribe('smoke_alert')


# Callback fires when a published message is received.
def on_message(client, userdata, msg):
    print(msg)
    z = [float(x) for x in msg.payload.decode("utf-8").split(',')]
    if z[0] == 1.0:
        message = twilio_client.messages.create(body='Smoke/Fire
                                                Detected',from_= from_whatsapp_number,
                                                to= to_whatsapp_number)

        print(message.sid)


client = mqtt.Client()
client.on_connect = on_connect# Specify on_connect callback
client.on_message = on_message# Specify on_message callback
client.connect('localhost', 1883, 60)  # Connect to MQTT broker

# Processes MQTT network traffic, callbacks and reconnections. (Blocking)
client.loop_forever()
```

After successfully running the code, the Twilio message notification is sent to the registered mobile number of the house owner that there is the fire in the house and necessary action has to be taken.

Conclusion:

NVIDIA® Jetson Nano™ lets you bring incredible new capabilities to millions of small and power-efficient AI systems. It opens new worlds of embedded IoT applications. Jetson Nano is a powerful computer that lets run multiple codes for various applications. In this chapter, the authors have given a brief introduction to FoG and Cloud computing along with its architectures, characteristics, service models, and various models deployment. It also covers the role of FoG and Cloud computing in IoT applications. The authors have provided the detailed implementation steps for the Patient Monitoring system with Cloud, wherein the system monitors Oxygen saturation, Pulse, and Body Temperature. Authors have also implemented Home Security system such as Home Surveillance, Home Safety Lock, and Fire alert system by providing detailed steps. This Home Security system is implemented with Jetson Nano as a FoG node.

Exercise:

(1) Similar to Sect. 5.6.2.3 (Fire alert system), using the ESP32-CAM module, design a system that reads the temperature from the DHT22 sensor and updates it over the FoG in regular intervals of 30 minutes.
(2) Modify the system mentioned in Sect. 5.6.1 (Patient Monitoring System with Cloud) to work on Raspberry Pi instead of Jetson Nano.
(3) Referring to Sect. 5.6.1 (Patient Monitoring System with Cloud) interface a MAX30102 Heart Rate sensor with Jetson Nano to monitor the patient's heart rate continuously. If the heart rate deviates outside the normal range and send an alert message via Twilio Cloud.
(4) Interface ESP32-CAM module with 2 PIR sensors referring to Sect. 5.6.2.2 (Home Safety Lock). Also interface a light source through relay module. Write a Python program which can make the ESP32-CAM module to take inputs from both the PIR sensors. If both the PIR sensors detect movement, then send the signals indicating the same to Jetson Nano and turn on the light source. Further, send the message "Movement detected, Light On" through Twilio cloud to WhatsApp number.
(5) Modify the setup in Sect. 5.6.2.1 (Home Surveillance) to capture images in interval of 10 minutes if there is no movement detected by the PIR sensor.
(6) To enhance the Home Security, interface fingerprint sensor module with system.
(7) Increase the number of IoT nodes with smoke sensors in Fire Alert system to detect fire in multiple rooms.

References

Aazam M, Huh EN (2014) Fog computing and smart gateway based communication for cloud of things. In: Proceedings of the 2014 international conference on future internet of things cloud, FiCloud 2014, Barcelona, Spain, 27–29 August 2014, pp 464–470

Aazam M, Huh EN (2015) Fog computing micro datacenter based dynamic resource estimation and pricing model for IoT. In: Proceedings of international conference on advanced information network applications, AINA 2015, pp 687–694

Aazam M, Hung PP, Huh E (2014) Smart gateway based communication for cloud of things. In: Proceedings of the 2014 IEEE ninth international conference on intelligent sensors, sensor networks and information processing, Singapore, 21–24 April 2014, pp 1–6

Arasaratnam O (2011) Introduction to cloud computing. In: Halpert B (ed) Auditing cloud computing, a security and privacy guide. Wiley, Hoboken, NJ, pp 1–13

Atlam HF, Walters RJ, Wills GB (2018) Fog computing and the internet of things: a review. Big Data Cogn Comput 2:10. https://doi.org/10.3390/bdcc2020010

Cha H-J, Yang H-K, Song Y-J (2018) A study on the design of fog computing architecture using sensor networks. Sensors 18:3633. https://doi.org/10.3390/s18113633

Chandrasekaran K (2015) Essentials of cloud computing. http://www.crcnetbase.com/isbn/9781482205442

Chiang M, Zhang T (2016) Fog and IoT: an overview of research opportunities. IEEE Internet Things J 3(6):854–864

Dastjerdi AV, Gupta H, Calheiros RN, Ghosh SK, Buyya R (2016) Fog computing: principles, architectures, and applications. In: Buyya R, Vahid Dastjerdi A (eds) Internet of things. Morgan Kaufmann, pp 61–75. https://doi.org/10.1016/B978-0-12-805395-9.00004-6

Evans D (2011) The internet of things how the next evolution of the internet is changing everything. https://www.cisco.com/c/dam/en_us/about/ac79/docs/innov/IoT_IBSG_0411FINAL.pdf. Accessed 13 Nov 2021

Fog Computing and Its Role in the Internet of Things. https://readwrite.com/2020/10/30/fog-computing-and-its-role-in-the-internet-of-things/

ITCandor (2018) Distribution of cloud platform as a service (PaaS) market revenues worldwide from 2015 to June 2018, by vendor. https://www.statista.com/statistics/540521/worldwidecloud-platform-revenue-share-by-vendor/

Iyer B, Henderson JC (2010) Preparing for the future: understanding the seven capabilities of cloud computing. MIS Q Exec 9(2):117–131

Khan S, Parkinson S, Qin Y (2017) Fog computing security: a review of current applications and security solutions. J Cloud Comput. 6(1):19

Kumar N, DuPree L (2011) Protection and privacy of information assets in the cloud. In: Halpert B (ed) Auditing cloud computing, a security and privacy guide. Wiley, Hoboken, NJ, pp 97–128

Liu Y, Fieldsend JE, Min G (2017) A framework of fog computing: architecture challenges and optimization. IEEE Access 4:1–10

Marques B, Machado I, Sena A, Castro MC (2017) A communication protocol for fog computing based on network coding applied to wireless sensors. In: Proceedings of the 2017 IEEE International symposium on high performance computer architecture, Vosendorf, Austria, 24–28 February 2017, pp 109–114

Marston S, Li Z, Bandyopadhyay S, Zhang J, Ghalsasi A (2011) Cloud computing—the business perspective. Decis Support Syst 51(1):176–189

Mell P, Grance T (2011) The NIST definition of cloud computing. National Institute of Standards and Technology, NIST Special Publication (SP), pp 800–145. https://doi.org/10.6028/NIST.SP.800-145

Mukherjee M, Shu L, Wang D (2018) Survey of fog computing: fundamental, network applications, and research challenges. IEEE Commun Surv Tutor

Muntjir M, Rahul M, Alhumyani HA (2017) An analysis of internet of things (IoT): novel architectures, modern applications, security aspects and future scope with latest case studies. Int J Eng Res Technol 6:422–447

Ara R, Rahim MA, Roy S, Prodhan UK (2020) Cloud computing: architecture, services, deployment models, storage, benefits and challenges. Int J Trend Sci Res Dev 4(4):837–842

Schneider S, Sunyaev A (2016) Determinant factors of cloud-sourcing decisions: reflecting on the IT outsourcing literature in the era of cloud computing. J Inf Technol 31(1):1–31

Shi Y, Ding G, Wang H, Roman HE, Lu S (2015) The fog computing service for healthcare. In: Proceedings of the 2015 2nd international symposium on future information and communication technologies for ubiquitous healthcare (Ubi-HealthTech), Beijing, China, 28–30 May 2015, pp 1–5

Sunyaev A (ed) (2020) "Cloud computing," in internet computing: principles of distributed systems and emerging internet-based technologies. Springer International Publishing, Cham, pp 195–236. https://doi.org/10.1007/978-3-030-34957-8_7

Taneja M, Davy A (2016) Resource aware placement of data analytics platform in fog computing. Proc Comput Sci 97:153–156. https://doi.org/10.1016/j.procs.2016.08.295

Tang B, Chen Z, Hefferman G, Wei T, He H, Yang Q (2015) A hierarchical distributed Fog computing architecture for big data analysis in smart cities. In: Proceedings of the ASE Big Data & Social Informatics 2015. ACM, p 28

Voorsluys W, Broberg J, Buyya R (2011) Introduction to cloud computing. In: Buyya R, Broberg J, Goscinski A (eds) Cloud computing. Wiley, Hoboken, NJ, pp 3–42

Chapter 6
Machine Learning (ML) in IoT with Jetson Nano

Abstract Machine Learning (ML) is a fast growing field and has been deployed in various IoT applications. When used in conjunction, they can deliver insights which are otherwise hidden in large data. One of the popular platforms used for ML in IoT is Jetson Nano SBC. It boasts of NVIDIA Maxwell architecture with 128CUDA cores, which make running ML algorithms efficient. This chapter lays the foundation for ML in a systematic manner by covering concepts such as what is AI, Deep Learning (DL) and ML. To provide a hands-on approach on ML, the authors have carefully curated three applications in the field namely, Pattern Recognition using ML with Cloud, Object Classification using ML with FoG, and Prediction of unknown Glucose concentration using ML at Edge. After completion of the chapter, the reader will have a firm grasp on the skills required to implement ML applications on the IoT platform.

Keywords Artificial intelligence · Machine learning · Pattern recognition · Object classification · Prediction · Edge

6.1 What is AI?

Artificial Intelligence (AI) was born in the 1950s, when Alan Turing in his pioneering paper titled "Computing Machinery and Intelligence (1950)" posed the question, "whether computers could be made to think?", a question whose answers is still being explored today. At its core, AI is the branch of computer science that seeks to answer Turing's question in the affirmative. It is an attempt to recreate or reproduce human intellect in machines. As such, AI is a general field that encompasses ML and DL as depicted in Fig. 6.1.

Many scientists have believed for a long time, that human-level AI could be achieved. The programmers tried to code a sufficiently extensive set of explicit rules for manipulating knowledge. This approach is called symbolic AI and was dominant

The original version of this chapter was revised. Series editor name has been removed from online version of this chapter. The correction to this chapter is available at https://doi.org/10.1007/978-981-19-9466-1_7

J. S. Parab et al., *Python Programming Recipes for IoT Applications*, Transactions on Computer Systems and Networks, https://doi.org/10.1007/978-981-19-9466-1_6

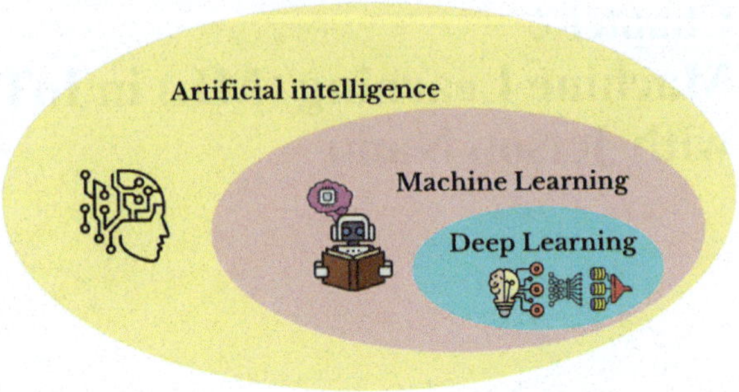

Fig. 6.1 Basic concept of AI

from the 1950s to the 1980s (Levesque 2017). The experts system boom in the 1980s is proved to be the golden period of symbolic AI.

Although symbolic AI successfully addressed well-defined logical problems like chess, it failed to devise clear rules for solving more complicated fuzzy problems like image classification, speech recognition, and language translation. To take the role of symbolic AI, two new approaches emerged: ML and DL.

6.2 Concepts of Machine Learning (ML) and Deep Learning (DL)

Machine Learning

ML is a marriage of two fields, namely, computer science and statistics which can be used to solve fuzzy tasks. The original question posed by Alan Turing opens the door to a new programming paradigm. In conventional programming and symbolic AI, the output is generated based on the data provided to the input rules (program). With ML, input data, as well as the expected output from the data is given, which produces a set of rules (program). These rules can then be applied to new data to produce the original output. Both the approaches are depicted in Fig. 6.2. Supervised learning, unsupervised learning, and reinforcement learning are the three most commonly prevalent forms of ML algorithms.

Supervised Learning:

In supervised learning, train the machine using data that is well "labeled". Supervised learning is used for problems related to regression and classification. The most commonly used supervised learning algorithms are Support-Vector Machines (SVM), Linear Regression, Logistic Regression, Naive Bayes, Linear Discriminant Analysis (LDA), Decision Trees (DT), K-Nearest Neighbor algorithm (KNN), Neural Networks-Multilayer Perceptron (MLP), etc. (Algorithms and of Machine Learning: Prediction of Brand Loyalty. 2019).

Unsupervised Learning:

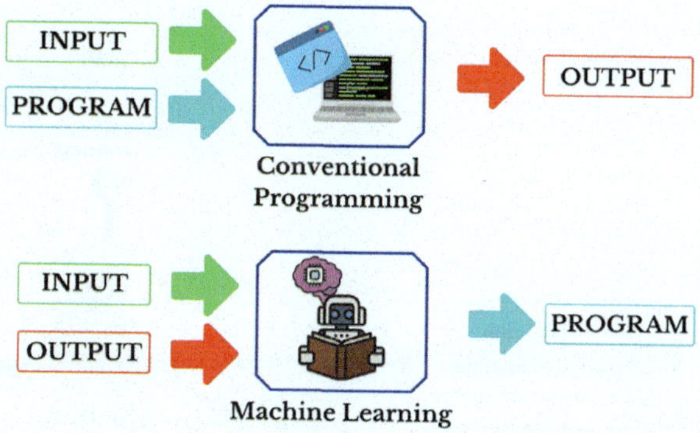

Fig. 6.2 Conventional programming versus ML

Unsupervised learning is a technique of ML in which the algorithm is not given any labels or scores for training the data. Unsupervised learning is used for dimension reduction and clustering. Few commonly used unsupervised learning algorithms are K-means clustering, KNN, Hierarchal clustering, Anomaly detection, Neural Networks (NN), Principle Component Analysis (PCA), Independent Component Analysis (ICA), Apriori algorithm, etc.

Reinforcement Learning:

Reinforcement Learning is an ML technique in which a model learns to perform the actions based on the feedback. For each accurate action, the model gets positive feedback, and for each inaccurate action, the model gets a negative feedback. Here, the model learns automatically using feedbacks without any labeled data, unlike in supervised learning. Few commonly used reinforcement methods are Monte Carlo, Q-learning, State–Action–Reward–State–Action (SARSA), Deep Q Network (DQN), Deep Deterministic Policy Gradient (DDPG), Q-Learning with Normalized Advantage Functions (NAF), etc. Fig. 6.3 represents the ML techniques in pictorial form.

Deep Learning (DL):

DL is a subfield of ML to solve sophisticated problems by extracting the hidden information by using the deeper layers. The high performance of DL can be realized when trained on the huge amount of data, which is not possible with conventional ML.

DL algorithms analyze data with a logic similar to how a human would conclude. DL applications do this by employing a layered structure of algorithms known as an Artificial Neural Network (ANN). The architecture of ANN is inspired by the biological neural network of the human brain, resulting in a learning process that is significantly superior to that of ordinary ML models.

Fig. 6.4 depicts a rudimentary ANN architecture. The left most layer is known as the 'Input layer', while the right most layer is known as the 'Output layer'. The layers between the input and output layer are referred to as 'Hidden layers'. The weights of the hidden layers are tuned so that the network is able to achieve the

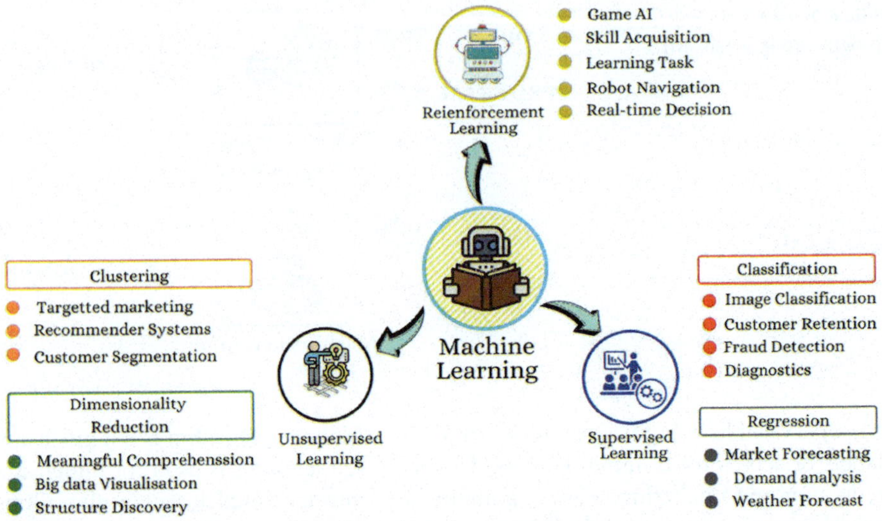

Fig. 6.3 ML techniques

desired goal. Deeper the network, the more hidden layers it has between the input and output levels. In general, a Deep Neural Network (DNN) is an ANN with one or more hidden layers.

DL has made drastic strides for applications involving huge data and high computing power to process these data. However, with the advent of cloud computing infrastructure and high-performance Graphics Processing Units (GPUs), the time required to train a DL network may be lowered.

However, the introduction of Transfer Learning (TL), i.e. the use of pre-trained models, is perhaps one of the most significant advancements in the field of DL. The

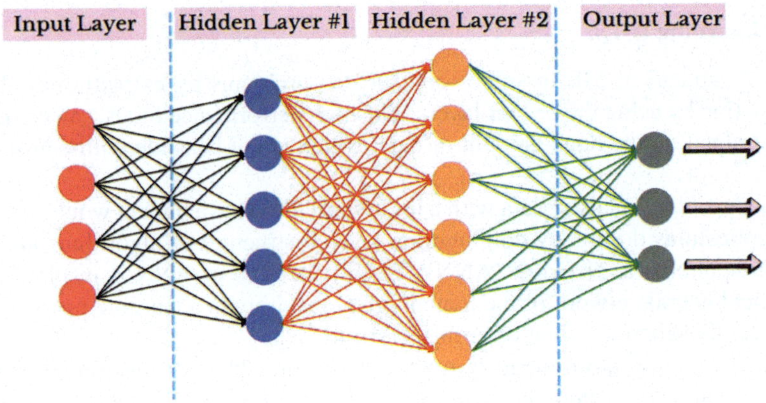

Fig. 6.4 A simple ANN with Input layer, two Hidden layers and Output layer

advent of TL has eliminated the requirement of huge data sets which were previously required for ANN to get better performance.

DL algorithms are very popular and most prominent among them are Convolutional Neural Networks (CNNs), Long Short Term Memory Networks (LSTMs), Recurrent Neural Networks (RNNs), Generative Adversarial Networks (GANs), Radial Basis Function Networks (RBFNs), Self Organizing Maps (SOMs), Deep Belief Networks (DBNs), Restricted Boltzmann Machines (RBMs), and Autoencoders.

Real-world DL applications are part of our everyday lives, but in most cases, they are so well-integrated into goods and services that consumers are clueless about the massive data processing that is going on behind the scenes. Examples include law enforcement, healthcare, customer service, financial services, and so on.

6.3 Pattern Recognition Using ML with Cloud

Pattern Recognition has piqued the interest of researchers all around the world. A pattern can be physically detected or computationally observed by using ML algorithms.

Pattern recognition systems can recognize familiar patterns quickly and accurately. Here, the labeled training data is used. Pattern recognition applications include image processing, voice, fingerprint identification, aerial picture interpretation, optical character recognition (OCR) and even medical imaging.

To understand the pattern recognition application, the authors have implemented classification of different shapes. There are in particular two classes: Quadrilaterals and Triangles. During the training phase, authors have fed the algorithm with 123 images of each class. For recognition, the model takes the image from a specified location and provides it as an input to the trained SVM ML model. ML model then predicts the class to which the image belongs and sends the message on WhatsApp via 'Twilio' cloud. The conceptual diagram for pattern recognition with the cloud is shown in Fig. 6.5. The SVM achieved pattern recognition accuracy of more than 90%.

Implementation Steps for Pattern Recognition with Cloud

This section covers implementation steps for Pattern Recognition using the SVM ML algorithm on the Jetson Nano SBC platform. After successful recognition, a WhatsApp message will be sent to a registered mobile number using the Twilio cloud platform.

SVM Model training and validation: Implementing the aforementioned application requires configuring Jetson Nano SBC by following the below steps.

Step 1: Update and Upgrade the platform using the following command as a root user.

```
sudo apt-get update
```

Fig. 6.5 Conceptual diagram for pattern recognition

```
sudo apt-get upgrade
```

Step 2: Install essential libraries/repositories.

The following libraries such as *'pip'* to get the required libraries, *'open-CV'* to manipulate images, *'sklearn'* for ML algorithm, and *'pickle'* for storing the trained model are to be installed by specified commands in Jetson Terminal.

```
sudo apt install python3-pip
sudo pip install opencv-python
sudo pip install -U scikit-learn
'sudo pip install pickle-mixin
```

Step 3: The next step is to arrange project directories in an orderly manner:

Create a directory named 'Project' in desired location (Authors have chosen Desktop as the desired location to harbor our project files). Open the 'Project' directory and within the Project directory, create a subdirectory named 'Pattern_recog'. This directory shall hold the dataset, trained model, and program for the aforementioned application.

Step 4: Configure Twilio on Jetson Nano SBC as described in Sect. 5.6.2.1 of Chap. 5.

Step 5: Enter the 'Pattern_recog' directory that was preciously created.

Step 6: Within the 'Pattern_recog directory', create 3 subdirectories and name them as 'Test', 'Triangle', and 'Quadrilateral', respectively.

Step 7: Save images of Quadrilaterals in 'Quadrilateral', Triangles in 'Triangle' and test images to be recognized in 'Test' directory, respectively.

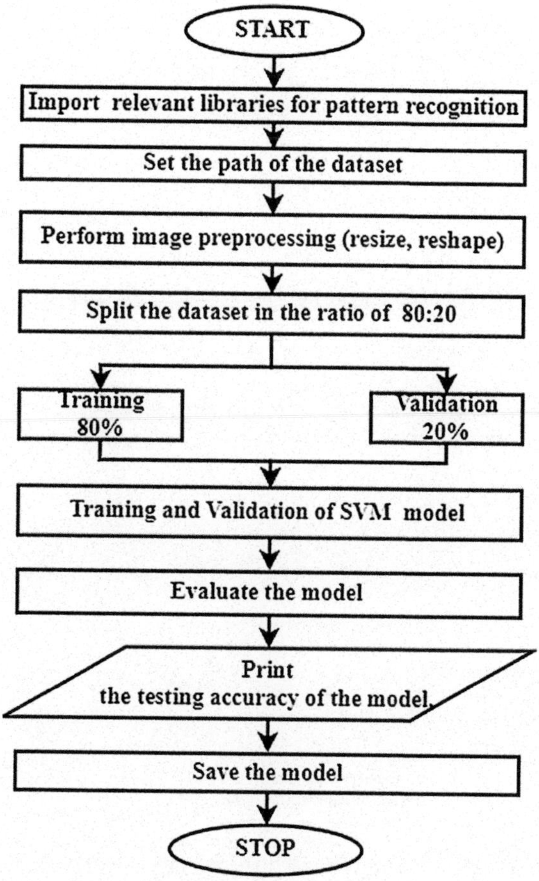

Fig. 6.6 Flow chart for the SVM model training and validation

Step 8: Save the Python code as *'model.py'* in 'Pattern_recog' directory and execute. This code builds SVM model and performs the model training and validation. The accuracy score of the same is displayed. After execution, the trained model pickle file having extension '.sav' is saved in the 'pattern_recog' directory. The flow chart for SVM model training and validation is given in Fig. 6.6.

```
"""This is a program for Training the model to detect
Triangles and Quadrilaterals."""
#'model.py'
# Importing various libraries needed.
import os
import numpy as np
import time
import cv2
import pickle
from sklearn.model_selection import train_test_split
from sklearn.svm import SVC
from sklearn.metrics import accuracy_score

# Accessing the data from the dataset folder.

file = '/home/jetson/Desktop/Project/Pattern_recog'
categories = ['Quadrilateral', 'Triangle']

data = []
features = []
labels = []

# Data preparation


for category in categories:
    path = os.path.join(file, category)
    label = categories.index(category)

    for img in os.listdir(path):
        image_path = os.path.join(path, img)
        image_shape = cv2.imread(image_path, 0)
        try:
            #Resizing images to 100x100 size and reshape.
            image_shape = cv2.resize(image_shape, (100, 100))
            image = np.array(image_shape).flatten()
            data.append([image, label])
        except Exception as e:
            pass

for feature, label in data:
    features.append(feature)
    labels.append(label)

# Splitting total dataset into training and testing set with 80:20 ratio.
X_train, X_test, Y_train, Y_test = train_test_split(features, labels, test_size=0.20)
#Fitting  and training the model on training set.
model = SVC(C=1, kernel='poly', gamma='auto',degree=4,coef0=1.5)
model.fit(X_train, Y_train)

# Prediction
Y_pred = model.predict(X_test)

# Printing Test acurracy of model.
accuracy = accuracy_score(Y_test, Y_pred)
print('Accuracy of the model is : ', accuracy)
#Saving the trained model as a .sav file using Pickle.
pick = open('model.sav','wb')
pickle.dump(model,pick)
pick.close()
```

Identification of unknown (Test) Image: This section allows user to test unknown shape (quadrilateral or triangle) using above trained SVM model.

Save the Python code as 'main.py' in 'Pattern_recog' directory and execute the code. This code takes unknown image which is saved in 'Test' folder for identification and sends the result to WhatsApp via Twilio cloud. The detailed flow chart for the

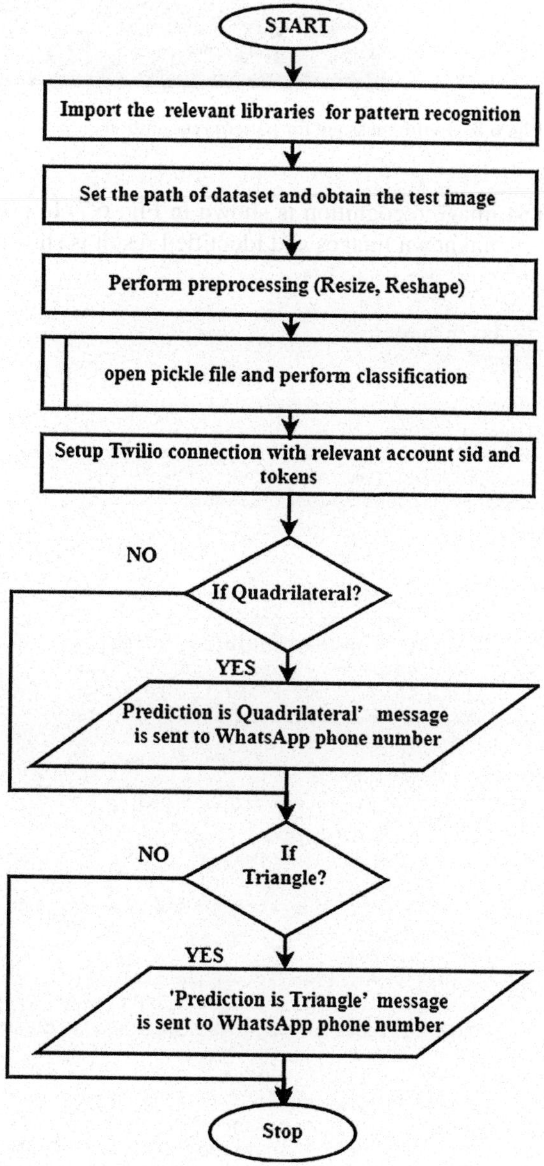

Fig. 6.7 Flow chart for the test image recognition

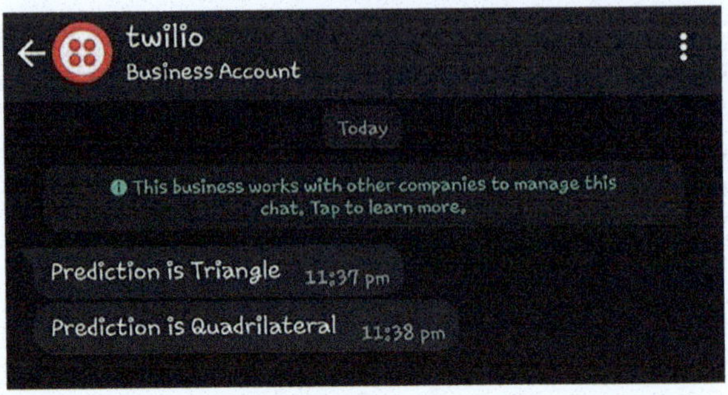

Fig. 6.8 Twilio message for pattern recognition

test image recognition is shown in Fig. 6.7. In present scenario, the authors tested two unknown images and identified result is shown in Fig. 6.8.

```
# This is main.py
import os
import numpy as np
import cv2
import pickle
from sklearn.model_selection import train_test_split
from sklearn.svm import SVC
from twilio.rest import Client

test = []

path = '/home/jetson/Desktop/Project/Pattern_recog/Test'
for img in os.listdir(path):
    img_path = os.path.join(path, img)
    image_shape = cv2.imread(img_path, 0)
    image_shape = cv2.resize(image_shape , (100, 100))
    image = np.array(image_shape).flatten()
    image = image.reshape(1,-1)


pick = open('model.sav','rb')
model = pickle.load(pick)
result = model.predict(image)

client = Client('your_account_sid', 'your_auth_token')

if result == [0]:
    print('Prediction is : Quadrilateral')
    message = client.messages.create(body='Prediction is 
    Quadrilateral',from_='whatsapp:+14155238886',to='whatsapp:+917057504996')

if result == [1]:
    print('Prediction is : Triangle')
    message = client.messages.create(body='Prediction is 
    Triangle',from_='whatsapp:+14155238886',to='whatsapp:+91XXXXXXXXXX')

print(message.sid)
```

6.4 Object Classification Using ML with FoG

Object detection combines classification and localization to determine whether items are present in an image or video and where they are located. The categorizing of an item based on previously determined classes or types is known as object classification. There are numerous object classification algorithms such as Logistic Regression, Naive Bayes, KNN, DT, SVM etc.

In this example authors have dealt with classifying different board images, i.e. Raspberry Pi, Jetson Nano, and Pyboard with help of the KNN algorithm, and is a three-class problem. The Jetson Nano is configured as a FoG node and ESP32-CAM as IoT node. FoG node receives a test image captured by ESP32-CAM module via MQTT and saves it to the *'Test'* directory on Jetson Nano. The KNN model takes the image from the 'Test' directory and applied to a trained KNN model. The model is trained with 600 images. KNN model then predicts the class to which the image belongs and sends the message on WhatsApp via Twilio cloud. The conceptual diagram for the Object Classification system is depicted in Fig. 6.9. The achieved classification accuracy is 100%.

Fig. 6.9 Conceptual diagram of object classification system

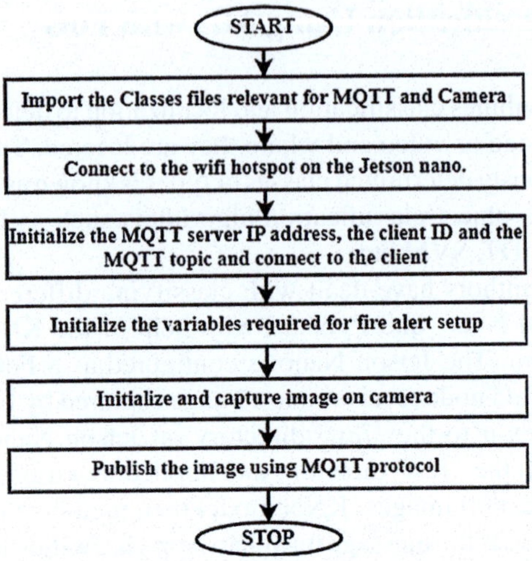

Fig. 6.10 Flow chart for configuring IoT node for image capture

Implementation Steps for Object Classification with FoG:

Step 1: Configuring IoT Node for Capturing the Image

Here, follow the IoT node configuration steps for capturing and publishing test image using MQTT protocol as given in Sect. 5.6.2.1 of Chap. 5. The flowchart for configuring the IoT node for capturing images is given in Fig. 6.10.

Step 2: Save the below code in ESP32-CAM as *'main.py'* using Thonny IDE and execute the same by clicking the run button to send the image via MQTT.

```
#main.py
from machine import Pin        #importing classes
from umqtt.simple import MQTTClient
import time
import os
import camera

# make sure wifi hotspot is enabled on jetson nano SBC(FoG Node)
import network
station = network.WLAN(network.STA_IF)
station.active(True)
station.connect("jetson-desktop", "11111111")

SERVER = '10.42.0.1'  # MQTT Server Address (Change to the IP address of your Pi)
CLIENT_ID = 'esp32-camera'
TOPIC = b'Image'

# Connect to MQTT broker
c = MQTTClient(CLIENT_ID, SERVER)
c.connect()

camera.init(0, format=camera.JPEG)

buf = camera.capture()
c.publish(mqtt_config[TOPIC], buf)
time.sleep_ms(100)
```

Step 3: Creating project directory on Jetson Nano (FoG node).

Create a directory named 'Project' in desired location (Authors have chosen Desktop as a desired location to harbor project files). Open the Project directory and create a subdirectory named as 'Object_classify'. This directory shall hold dataset, pickle file of trained KNN model, and Python code.

Step 4: Configure Twilio and MQTT on Jetson Nano as described in Sect. 5.6.2.1 of Chap. 5.

Step 5: Create 4 subdirectories under 'Object_classify' directory and name them as 'Test', 'Raspberry_pi', 'Jetson_nano', and 'Py_board' to save the respective images. The flow chart for the training process is given in Fig. 6.11.

Step 6: Save the following code as 'model.py' in 'Object_classify' directory which is created in Jetson Nano.

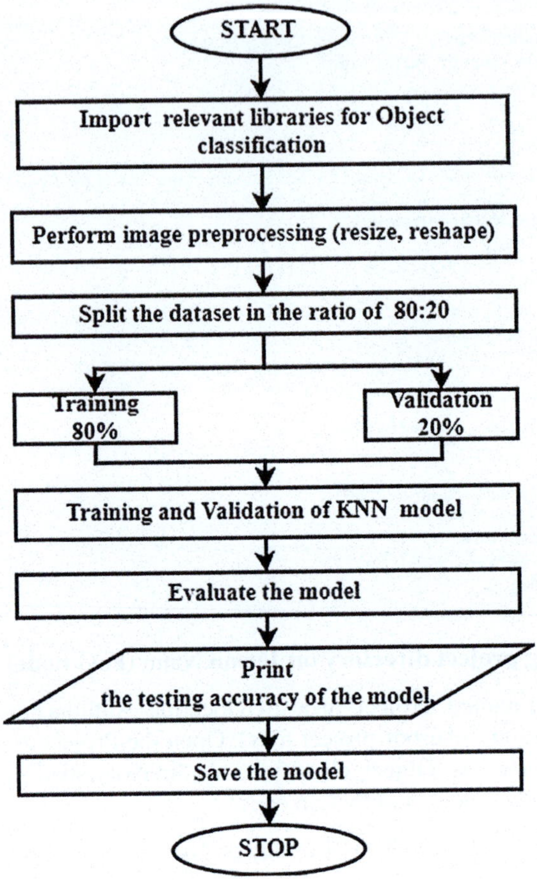

Fig. 6.11 Flow chart for training the KNN model

```
"""This is a program tarins the model for object classification. There
are three classes, namely Raspberry Pi, Py Board and Jetson Nano"""

# Importing various libraries needed.
import os
import numpy as np
import time
import cv2
import pickle
from sklearn.model_selection import train_test_split
from sklearn.neighbors import KNeighborsClassifier
from sklearn.metrics import accuracy_score

# Accessing the data from the dataset folder.

file = '/home/jetson/Desktop/Project/Object_classify'
categories = ['Py_board', 'Raspberry_pi','Jetson_nano']

data = []
features = []
labels = []

# Data preparation
```

```
for category in categories:
    path = os.path.join(file, category)
    label = categories.index(category)

    for img in os.listdir(path):
       image_path = os.path.join(path, img)
       image_shape = cv2.imread(image_path, 0)
       try:
           #Resizing images to 100x100 size and reshape.
           image_shape = cv2.resize(image_shape, (100, 100))
           image = np.array(image_shape).flatten()
           data.append([image, label])
       except Exception as e:
           pass

for feature, label in data:
    features.append(feature)
    labels.append(label)

# Splitting total dataset into training and testing set in 80 : 20 ratio.


X_train, X_test, Y_train, Y_test = train_test_split(features, labels, test_size=0.20)


#Fitting  and training the model on training set.

model = KNeighborsClassifier(n_neighbors=3)
model.fit(X_train, Y_train)


# Prediction
Y_pred = model.predict(X_test)

# Printing Test acurracy of model.
accuracy = accuracy_score(Y_test, Y_pred)

#Saving the trained model as a .sav file using Pickle.

pick = open('model_obj.sav','wb')
pickle.dump(model,pick)
pick.close()
```

Step 7: Right-click inside the directory and click on 'open in terminal' (To open terminal in working directory). Execute the code by typing 'sudo python3 model.py'. After execution, trained model having pickle file with extension '.sav' will be saved in the 'Object_classify' directory.

Step 6: Save the following code as 'main.py' in 'Object_classify' directory and execute. This code receives an image from IoT node (ESP-32CAM) through MQTT protocol. This received image is then saved in 'Test' directory. The flowchart for image classification is given in Fig. 6.12. The image is then classified and the classification result is sent on the WhatsApp via Twilio cloud as shown in Fig. 6.13.

Fig. 6.12 Flow chart for classification

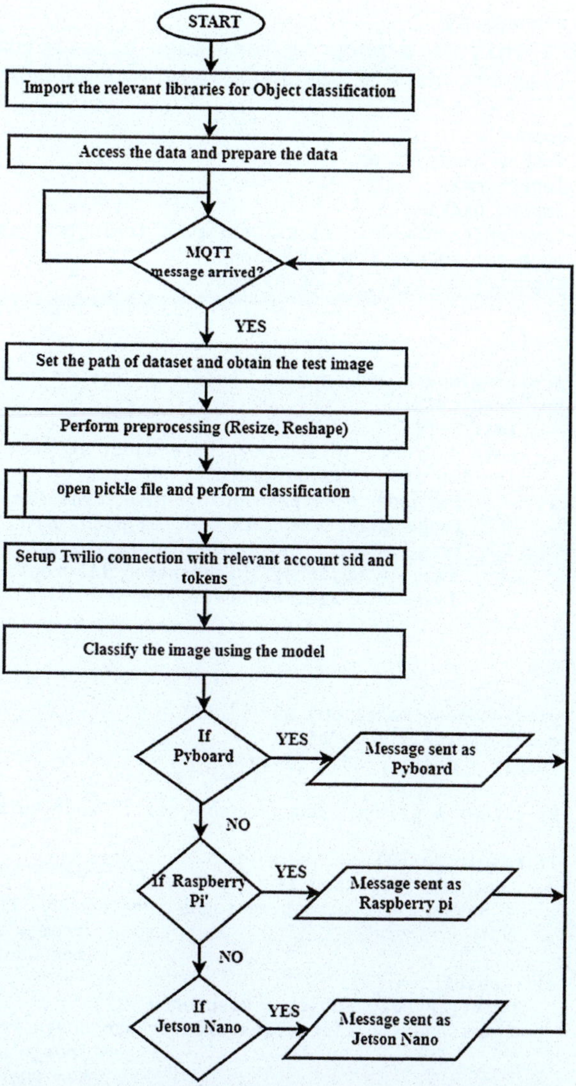

Fig. 6.13 Twilio cloud message for object classification

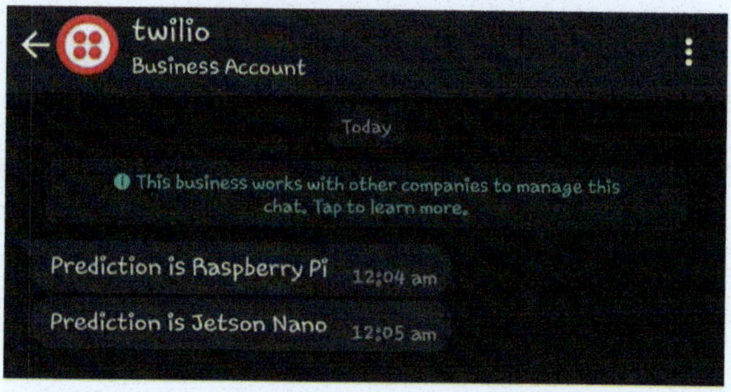

```
# main.py
"""This is a program for image classification. There are three
classes, namely Raspberry Pi, Py Board and Jetson Nano. """

import os
import numpy as np
import cv2
import pickle
from sklearn.model_selection import train_test_split
from twilio.rest import Client
import paho.mqtt.client as mqtt



#calling model and sending  predicted result on whatsapp
def predict():
    test = []
    path = '/home/jetson/Desktop/Project /Object_classify/Test'
    for img in os.listdir(path):
        img_path = os.path.join(path, img)
        image_shape = cv2.imread(img_path, 0)
        image_shape = cv2.resize(image_shape , (100, 100))
         image = np.array(image_shape).flatten()
         image = image.reshape(1,-1)



pick = open('model_obj.sav','rb')
model = pickle.load(pick)
result = model.predict(image)

#t_client = Client('your_account_sid', 'your_auth_token')

if result == [0]:
     print('Prediction is : Py Board')
    t_message = t_client.messages.create(body='Prediction is Py Board',
                                         from_='whatsapp:+14155238886',
                                         to='whatsapp:+91XXXXXXXXXX')

elif result == [1]:
    print('Prediction is : Raspberry Pi')
    t_message = t_client.messages.create(body='Prediction is Raspberry Pi',
                                         from_='whatsapp:+14155238886',
                                         to='whatsapp:+91XXXXXXXXXX')

elif result == [2]:
    print('Prediction is : Jetson Nano')
        t_message = t_client.messages.create(body='Prediction is JesonNano',
                                             from_='whatsapp:+14155238886',
                                             to='whatsapp:+91XXXXXXXXXX')

    print("message sent")
```

```
def on_connect(client, userdata, flags, rc):
       print('Connected with result code {0}'.format(rc))
      # Subscribe (or renew if reconnect).
      client.subscribe('glucose')

def on_message(client, userdata, msg):
    print(msg)
    print("Message received-> " + msg.topic + " " + str(msg.payload)) f =
    open('test/'+'Test_image.jpg', "wb")
    f.write(msg.payload)
    f.close()
    print("Image received and saved!")
    predict()

client = mqtt.Client()
client.on_connect = on_connect# Specify on_connect callback
client.on_message = on_message# Specify on_message callback
client.connect('localhost', 1883, 60)  # Connect to MQTT broker.
client.loop_forever()
```

6.5 Prediction of Unknown Glucose Concentration Using ML at EDGE

For longevity of life, one has to monitor body parameters regularly which also helps to avoid serious complications caused by various diseases such as diabetes, cardiovascular diseases, etc. Diabetes is a serious condition that is characterized by high blood glucose levels. If this level remains high for a long duration it leads to several health problems such as kidney failure, stroke, amputation of limbs, heart attack, blindness, etc. Hence, it is of utmost importance to monitor blood glucose regularly for diabetes patients. There are Glucose monitoring meters available in markets, but one needs to put a drop of blood by pricking the finger. Since it involves pricking, it causes pain and can lead to infection. To circumvent this problem, a non-invasive method is required such as optical absorption signatures from human tissues and presently, an extensive amount of research is being carried out in this particular area.

In this application, Jetson Nano SBC acts as an Edge node on which the Multilayer perceptron (MLP) model is deployed to accurately predict the glucose concentration. Edge computing is a distributed framework wherein client data is processed as near to the source as feasible at the network's perimeter. To consider the data set for prediction from the human blood sample one requires ethical clearance. Therefore, we have worked on laboratory samples instead of human tissues. The samples are prepared by mixing five major blood constituents as glucose, ascorbate, alanine, lactate, and urea with a physiologically relevant concentration in the normal range. Authors have created database of 57 samples having absorption signatures at three wavelengths, namely, 2246 nm, 2280 nm, and 2308 nm using Spectrophotometer Jasco V770.

MLP Model deployed on Edge is trained with 45 samples and 12 samples are used for validating the model, i.e. database split ratio is of 80:20. The MLP model uses 'RELU' activation function, maximum iteration of 30,000 and MinMaxscaler() normalization function is used for data standardization. The coefficient of determination(R2) for glucose prediction is 0.99.

Implementation Steps for Glucose Prediction at EDGE

This section covers implementation steps for Glucose prediction using MLP algorithm on Jetson Nano SBC platform acting as Edge.

Implementing the aforementioned application requires configuring Jetson Nano SBC by following the below steps.

Step 1: Update and Upgrade the platform using the following command as a root user.

```
sudo apt-get updat e.

sudo apt-get upgrade
```

Step 2: Install essential libraries/repositories.

The following libraries are to be installed by specified commands in Jetson Terminal.

```
sudo apt install python3-pi p

sudo pip install -U scikit-learn
sudo pip install pickle-mixin
sudo pip3 install pandas
```

Step 3:
Create a directory named 'Project' in your desired location (Authors have chosen Desktop). Open the 'Project' directory and create a subdirectory named as 'Glucose_pred'. This directory shall hold dataset, trained model, and program for the aforementioned application.

Step 4: Navigate to 'Glucose_pred' directory that was preciously created.

Step 5: Save the following code as 'model.py' in 'Glucose_pred' directory. The flow chart for glucose prediction is given in Fig. 6.14.

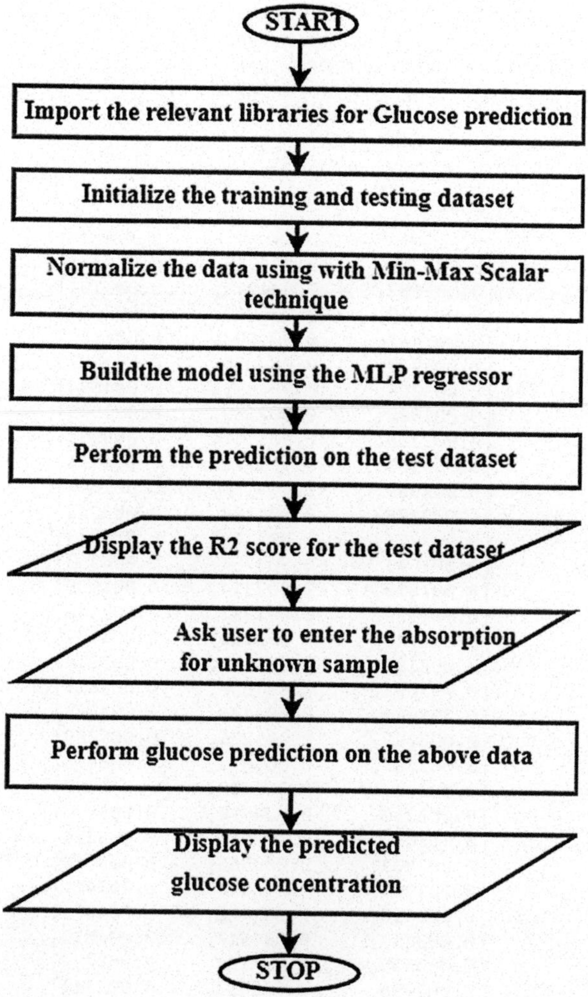

Fig. 6.14 Flow chart for glucose prediction

```
# MLP regressor code for Glucose Prediction at FoG
from sklearn.neural_network import MLPRegressor
from sklearn.model_selection import train_test_split
from sklearn.preprocessing import MinMaxScaler
from sklearn.metrics import r2_score
import pandas as pd
import numpy as np

"""Initialisedatset """

x_train=np.array([[0.003509,    0.004917,    0.004881],
                  [0.003682,    0.004913,    0.005101],
                  [0.002919,    0.004060,    0.003594],
                  [0.003482,    0.004803,    0.004826],
                  [0.003272,    0.004080,    0.003667],
                  [0.003396,    0.004575,    0.004694],
                  [0.001902,    0.002689,    0.002905],
                  [0.003768,    0.005141,    0.005233],
                  [0.003303,    0.004775,    0.004971],
                  [0.002589,    0.003133,    0.002664],
                  [0.003178,    0.004284,    0.003945],
                  [0.002221,    0.003144,    0.003373],
                  [0.003671,    0.004764,    0.004263],
                  [0.001810,    0.002314,    0.001939],
                  [0.002529,    0.002902,    0.002548],
                  [0.001720,    0.002090,    0.001807],
                  [0.002188,    0.003027,    0.003312],
                  [0.003092,    0.004060,    0.003813],
                  [0.003384,    0.004426,    0.003855],
                  [0.003795,    0.005256,    0.005289],
                  [0.003735,    0.005025,    0.005173],
                  [0.002562,    0.003019,    0.002608],
                  [0.003330,    0.004890,    0.005027],
                  [0.002010,    0.002420,    0.002214],
                  [0.002892,    0.003945,    0.003538],
                  [0.001961,    0.002920,    0.003021],
                  [0.003558,    0.004422,    0.004075],
                  [0.003217,    0.004547,    0.004839],
                  [0.003325,    0.004194,    0.003739],
                  [0.002096,    0.002653,    0.002347],
                  [0.002189,    0.002452,    0.002069],
                  [0.003017,    0.004437,    0.004564],
                  [0.002806,    0.003717,    0.003406],
                  [0.003449,    0.004686,    0.004765],
                  [0.002400,    0.003172,    0.003227],
                  [0.001777,    0.002197,    0.001879],
                  [0.003205,    0.004398,    0.004001],
                  [0.002653,    0.003393,    0.003574],
```

```
                   [0.003611,    0.004533,    0.004147],
                   [0.002930,    0.004209,    0.004432],
                   [0.002984,    0.004320,    0.004504],
                   [0.002600,    0.003282,    0.003502],
                   [0.002367,    0.003055,    0.003166],
                   [0.002859,    0.003828,    0.003478],
                   [0.002686,    0.003510,    0.003634]])

x_test=np.array([[0.002276,    0.002680,    0.002201],
                 [0.002063,    0.002536,    0.002286],
                 [0.003644,    0.004650,    0.004207],
                 [0.002302,    0.002795,    0.002256],
                 [0.002247,    0.003259,    0.003428],
                 [0.001837,    0.002429,    0.001995],
                 [0.002314,    0.002943,    0.003095],
                 [0.002243,    0.002563,    0.002140],
                 [0.003043,    0.004551,    0.004620],
                 [0.002427,    0.003286,    0.003282],
                 [0.002476,    0.002790,    0.002476],
                 [0.001848,    0.002577,    0.002833]])

y_train=np.array([[280],[70], [280],[200], [70], [70],[100], [200], [200], [280],
                  [200], [200], [280], [200], [100],[70], [100], [70],[280], [280],
                  [100], [200], [280], [70], [200], [280], [70], [70], [100], [200],
                  [70], [200], [70], [100], [200], [100], [280], [100], [100], [70],
                  [100], [70], [100], [100], [200]])

y_test=np.array([[200], [100], [200], [280], [280], [280], [70], [100], [280], [280],
                 [70], [70]])


# datanormalisation technique with Min-Max Scaler technique
scaler = MinMaxScaler()
scaler.fit(x_train)
x_train1 = scaler.transform(x_train)
x_test1 = scaler.transform(x_test)

model = MLPRegressor(hidden_layer_sizes=(11,8),activation="relu" ,random_state=1,
                     max_iter=30000).fit(x_train1, y_train)

y_pred=model.predict(x_test1)
print("The Score with ", (r2_score(y_pred, y_test)))

test=x_test1[0]

arr = np.array(x_test1[0])
actual=np.array(y_test)
# Convert 1D array to a 2D numpy array of 2 rows and 3 columns
test = np.reshape(arr, (1, 3))

actual[0]

predicted_all=model.predict(x_test1)
print(predicted_all)
```

```
predicted=model.predict(test)

W1 = float(input("Enter your glucose 1 wavelength value: "))
W2 = float(input("Enter your glucose 2 wavelength value: "))
W3 = float(input("Enter your glucose 3 wavelength value: "))

wave_values=np.array([[W1, W2, W3]])
wave_values = scaler.transform(wave_values)

predicted=model.predict(wave_values)
print ("Predicted glucose is ")
print(predicted)
```

Step 6: Run the above code using the following command in the Jetson Nano terminal (make sure the terminal is opened from the working directory).

```
sudo python3 model.py
```

After Successful execution the predicted results are shown on the Jestson terminal.

Conclusion:

ML is a fast growing field and has been deployed in various IoT applications. To start with, the authors have given the brief introduction of AI, DL, and ML. Authors have implemented three applications on Jetson Nano platform, namely, Pattern Recognition using ML with Cloud, Object Classification using ML with FoG and Prediction of unknown Glucose concentration using ML at Edge. The detailed implementation steps of all three applications are provided for ready prototyping for readers.

Exercise:

(1) With reference to the example in the Sect. 6.3 (Pattern Recognition using ML with Cloud), try running the program with different resolution of the images. E.g.: 200 × 200, 150 × 150, and 50 × 50 and see the change in prediction accuracy of the trained model.
   Hint: Change cv2.resize(image_shape, (100, 100)) to desired resolution.
(2) For the activity mentioned in the Sect. 6.3 (Pattern Recognition using ML with Cloud) examine the change in result when number of shapes in the dataset is increased from 2 to 3. E.g.: Quadrilateral, Triangle, and Circular/Elliptical. Also follow up with sending the result over to Twilio cloud.
(3) In the Sect. 6.4 (Object Classification using ML with FoG), explore the possibility of replacing the FoG node, i.e. Jetson Nano with Raspberry Pi.
(4) In the Sect. 6.4 (Object Classification using ML with FoG), try to create your own dataset and evaluate the performance with various ML models.
(5) Investigate how changing parameters such as C, degree and coefficient affects the accuracy of system mentioned in Sect. 6.3 (Pattern Recognition using ML with Cloud). Further determine the accuracy that can be achieved by using KNN classifiers.
(6) Build a dataset using your own glucose sensor and pass it to the above glucose prediction model and determine the prediction accuracy.

References

https://www.simplilearn.com/10-algorithms-machine-learning-engineers-need-to-know-article
https://machinelearningmastery.com/transfer-learning-for-deep-learning/
https://www.ibm.com/cloud/learn/deep-learning/
(2019) Supervised learning algorithms of machine learning: prediction of brand loyalty. IJITEE 8:3886–3889. https://doi.org/10.35940/ijitee.J9498.0981119
Levesque HJ (2017) Common sense, the turing test, and the quest for real AI: Reflections on natural and artificial intelligence. MIT Press

Correction to: Python Programming Recipes for IoT Applications

Correction to:
J. S. Parab et al., *Python Programming Recipes for IoT Applications*, Transactions on Computer Systems and Networks, https://doi.org/10.1007/978-981-19-9466-1

The original version of the book was inadvertently published with incorrect author names in online version.

The correction chapters and the book have been updated with these changes.

The updated versions of these Chapters can be found at
https://doi.org/10.1007/978-981-19-9466-1_1
https://doi.org/10.1007/978-981-19-9466-1_2
https://doi.org/10.1007/978-981-19-9466-1_6

J. S. Parab et al., *Python Programming Recipes for IoT Applications*, Transactions on Computer Systems and Networks, https://doi.org/10.1007/978-981-19-9466-1_7

The manufacturer's authorised representative in the EU is Springer Nature Customer Service Centre GmbH, Europaplatz 3, 69115 Heidelberg, Germany. If you have any concerns regarding our products, please contact ProductSafety@springernature.com

Printed and bound by CPI Group (UK) Ltd, Croydon, CR0 4YY
01/10/2026
02225877-0014